# Experimental Brain Research

W0232007

## 29th Annual General Meeting
## of the European Brain and Behaviour Society

*Special Editors*

**Nicole von Steinbüchel, Alexander Steffen, Marc Wittmann**

15–18 September 1997
Tutzing (Bavaria), Germany

Supplement to Volume 117 (1997) of Experimental Brain Research

Springer

ISBN 978-3-662-39400-7          ISBN 978-3-662-40459-1 (eBook)
DOI 10.1007/978-3-662-40459-1

Softcover reprint of the hardcover 1st edition 1997

## 1. General

**Manuscripts** should be typewritten, double-spaced throughout, with wide margins, and should be submitted in *triplicate* to one of the Section Editors (see inside cover) up to five referees can be suggested. Authors submitting diskettes are requested to follow the technical instructions printed in each issue of the journal. Papers should be clear and concise. They should be written in English. Correct language is the responsibility of the author(s). Author(s) should follow the typographical conventions of the journal as used in recent issues. Non-English-speaking authors should have their papers checked and amended by a native English speaker. English *or* American spelling should be used consistently throughout a manuscript.

**Manuscripts in electronic form.** Submission of electronic data is encouraged. Please do not transmit any such data to the publisher until your manuscript has been reviewed and accepted for publication. Please follow the instructions for the preparation of electronic manuscripts and digital images.

**The journal accepts *Research notes*. They can be submitted to any Section Editor and will usually be reviewed by the Section Editor and only one referee.**

The editors emphasize the virtue of brevity. If a paper is returned to the authors for revision, it must be back within 6 months.

To accelerate publication, **only one set of proofs** is sent to the authors. This shows the final layout of the paper as it will appear in the journal. It is therefore essential that manuscripts are submitted in their *final form*, ready for the printer, and that the *positions of figures and tables are indicated* in the margins. Proof reading should be *limited* to the correction of *typographical errors*. Any other changes involve time-consuming and expensive work, and *the costs will be charged to the author(s)*. If absolutely necessary, additions may be made at the end of the paper in a "Note added in proof".

Offprints may be ordered at cost price when the page proofs are returned.

**Supplementary electronic material.** Electronic supplementary material will be published free of charge on Springer Verlag's server http://science.springer.de if (a) submitted on diskettes or cartridges, together with the manuscript and a brief description including a list of files, and (b) accepted by the editors (details on http server).

## 2. Text

**Abstract.** Each article must be preceded by an abstract.

**Key words.** Following the abstract, the author(s) should provide up to five key words characterizing the scope of the paper.

**Headings.** Each *main heading* (Introduction, Material and methods, etc.) and

*subheading* should be placed on separate lines.

**Small print.** 'Materials and methods' and sections of lesser importance should be marked for small print. Small type is used to improve the layout of the paper.

**Genus and species names** and words to be emphasized should be underlined *singly* for *italics*.

**Measuring units.** Temperatures may be expressed in degrees Celsius; times apart from seconds (s), in minutes (min), hours (h), etc. Otherwise, the **International System of Units** (SI, Système International d'Unités) should be used wherever possible. (Consult, *e.g.*, the pamphlet "Metric Practice Guide, A Guide to the Use of SI – the International System of Units", Publ. E380-70, Am. Soc. for Testing and Materials, Philadelphia, Pennsylvania, USA). Other units may be given in parentheses when first appearing in the text. Biochemical terminology should follow that used by *J. Neurochem.* and *Eur. J.Biochem.*

**Footnotes** in the text should be numbered consecutively and placed at the bottom of the page to which they refer.

The list of **References** should only include publications cited in the text. The references should be cited in alphabetical order under the first author's name, listing all authors (surnames *followed by initials throughout*; do *not* use "and"), with the complete title, according to the following examples:

a) *Articles from journals:* name(s) and initials of all author(s), year in parentheses, full title, journal name as abbreviated in *Index Medicus*, volume followed by a colon, first and last page numbers.

Jacobson SG, Eames RA, McDonald WI (1979) Optic nerve fibre lesions in adult cats: pattern of recovery of spatial vision. ExpBrain Res 36: 491–508

b) *Books:* name(s) and initials of all author(s), year in parentheses, title, edition (ed), publisher, place of publishing.

Chan-Palay V (1977) Cerebellar dentate nucleus: organization, cytology, and transmitters. Springer, Berlin Heidelberg New York

c) *Multiauthor books:* name(s) and initials of all author(s), year in parentheses, title of the paper, name(s) and initials of all editor(s), title of the book, publisher, place of publishing, first and last page numbers.

Oscarsson O (1973) Functional organization of spinocerebellar paths. In: Iggo A (ed) Somatosensory system. Handbook of sensory physiology, vol II. Springer, Berlin Heidelberg New York, pp 339–380

## 3. Tables

Tables should be **numbered consecutively** with arabic numbers. **Footnotes in tables** should be indicated by lowercase super-

script letters, beginning with ᵃ in each table.

## 4. Illustrations

*Illustrations* should be limited to materials essential for the text. Previously published illustrations cannot normally be accepted. For line drawings please submit good-quality prints. Computer drawings are acceptable provided they are of comparable quality to line drawings. Lines and curves must be smooth.

*Color figures* can be published without charge if the editor agrees. Otherwise, the authors will be expected to make a contribution towards the extra costs (approx. DM 1200,– for the first and DM 600,– for each additional page).

*All figures* should be numbered consecutively and submitted separately from the text. The figures, including legends, should not exceed the print area of 17,6 cm × 23,6 cm or the column width of 8,6 cm × 23,6 cm. Figures should be marked on the back ([figure number and name(s) of author(s)]). Photo- or micrographs should be mounted together to save space without necessarily taking into consideration their numerical order. If plates are submitted, the figures must be mounted on regular bond paper, not on cardboard.

*Captions* must be brief, self-sufficient explanations of the illustrations in no more than four or five lines. Remarks like "For explanation, see text" should be avoided. The captions are part of the text and should be appended to it on a separate page. *Magnification* in micrographs should be indicated by *scale bars*, and not by quoting a factor.

## 5. Research notes

These should consist of not more than **five printed pages (2400 words)**, including references and **one or two figures at the most**; they should be preceded by a short abstract (not exceeding 250 words). This section enables authors to communicate interesting findings of their on going research to be published quickly. Research notes are not to be submitted for publication elsewhere as part of a full paper.

*Research Notes* are given priority over all other manuscripts for publication. Please therefore make a special effort to meet the deadline for return of your page proofs. If you encounter any problems, or if extensive proof corrections seem necessary, please inform us as early as possible by fax so that we can decide on an appropriate course of action. Should your corrected proofs not have reached us by three days before the issue is due to go to press, we may carry out the proofreading in order to avoid any delay.

## 6. Use of animals

Please refer to the advice in bold type on the inside front cover (page A2).

# 29ᵗʰ Annual General Meeting (15–18 September 1997) of the European Brain and Behaviour Society – EBBS

## Contents

| Section | Abstract-Number | Pages |
|---|---|---|
| **Invited Speakers** | Inv/1–Inv/5 | S1–S2 |
| **Symposia 1–14** | | |
| 1. Genetic model: How can they help to identify the neural basis of behaviour (organized by: M. Ammasari-Teule) | Sym1/1–Sym1/4 | S3–S4 |
| 2. Parietal stories (organized by: R. Caminiti) | Sym2/1–Sym2/4 | S4–S5 |
| 3. Colour and the brain (organized by: A. Cowey) | Sym3/1–Sym3/4 | S6–S7 |
| 4. Functional neuroimaging (organized by: R.S.J. Frackowiak) | Sym4/1–Sym4/5 | S7–S8 |
| 5. The formation of motor concepts (organized by: H.-J. Freund) | Sym5/1–Sym5/2 | S8–S9 |
| 6. Neuropeptides and behaviour (organized by: R.U. Hasenoehrl, R.K.W. Schwarting) | Sym6/1–Sym6/3 | S9–S10 |
| 7. Ocular motor control in human cerebrial cortex (organized by: W. Heide) | Sym7/1–Sym7/4 | S10–S11 |
| 8. Disorders of spatial perception in patients with neglect (organized by: H.-O. Karnath) | Sym8/1–Sym8/4 | S11–S12 |
| 9. Language processing and brain function (organized by: W.J.M. Levelt) | Sym9/1–Sym9/4 | S12–S13 |
| 10. Spatial memory (organized by: F. Schenk, R.G.M. Morris) | Sym10/1–Sym10/4 | S13–S14 |
| 11. Time and cognition (organized by: N. von Steinbuechel) | Sym11/1–Sym11/3 | S14–S15 |
| 12. Glutamatergic receptor subtypes: Functional implications in learning and memory (organized by: A. Ungerer) | Sym12/1–Sym12/4 | S15–S17 |
| 13. Infant's development of object perception and recognition behaviour and brain imaging (organized by: F. Vital-Durand) | Sym13/1–Sym13/6 | S17–S18 |
| 14 Short- and long-term modifications of circuitry in somatosensory cortex following changes in sensory experience (organized by: A. Wrobel, M. Armstrong-James) | Sym14/1–Sym14/4 | S18–S19 |
| **Poster Presentations** | | |
| DP – Disease and Pathology | DP/1–DP/19 | S21–S26 |
| ES – Emotion and Stress | ES/1–ES/14 | S27–S31 |
| FN – Functional Neuroanatomy and Functional Neurophysiology | FN/1–FN/8 | S32–S34 |
| LN – Learning and Memory | LM/1–LM29 | S35–S44 |
| LS – Language and Speech | LS/1–LS/12 | S45–S49 |
| MO – Movement | MO/1–MO/17 | S50–S55 |
| N – Neglect | N/1–N10 | S56–S58 |
| PP – Psychopharmacology | PP/1–PP/4 | S59–S60 |
| T – Time | T/1–T/14 | S61–S65 |
| V – Vision | V/1–V/20 | S66–S72 |
| **Author Index** | (according to abstract number) | I–III |

# Invited Speakers

## Inv1

PLACE CELLS AND PLACE NAVIGATION IN THE MOVING WORLD. *J. Bures. Institute of Physiology, Academy of Sciences, Videnska 1083, 14220 Prague 4, Czech Republic*

The ability of rats to navigate to goals not directly visible from the departure point is based on neural representation of the known environment in the form of cognitive maps implemented by hippocampus and perhaps by hippocampal place cells (PCs). The latter assumption can be tested by examining navigation during disruption of PC activity, e.g. by creating a conflict between egocentric (idiothetic) and allocentric (visual) orientation. Whereas during locomotion in stable world both allocentric and egocentric inputs yield the same information, an animal placed on a slowly rotating arena can perceive its own position either in the room frame, i.e. relative to the stable room, or in the arena frame, i.e. relative to the surface of the moving arena. A computerized tracking system recording the rat's position in either frame shows that during exploration of a rotating arena firing fields (FFs) of some PCs remain stable in the room frame, of others in the arena frame, but that most FFs disintegrate in both frames. The allocentric-egocentric conflict can be enhanced by the field clamp technique: The rat searching pellets in a ring shaped arena is always returned by rotation of the arena to the same allocentric position. In this case the conflict increases proportionally to the active locomotion of the rat. In 70% PCs 10 min of field clamp conditions caused transient disintegration or remapping of FFs. The influence of the above conditions on place navigation was investigated in the place avoidance task: a rat foraging on a stationary or rotating arena and receiving mild footshock upon entering an allocentrically defined region learned in a few min to avoid the prohibited zone. Extinction took more than 30 min when tested in light both on stable and rotating arena. During testing in darkness, avoidance was observed only after training on stable arena which has obviously led to formation of both allo- and egocentric representations. The latter was retrievable in darkness even on rotating arena where the avoided zone could be seen in the arena frame. Avoidance training on rotating arena produced an allocentric map alone which was not retrievable in darkness. In the place preference task the rat searching food on the entire surface of the arena is trained to visit of an allo- or egocentrically defined location, entering of which triggers delivery of another pellet to a random position on the floor. A well trained rat alternates between random search terminated by finding the pellet followed by rapid goal-directed run to the trigger area the finding of which delivers a pellet and starts a new search. Place preference on rotating arena is purely allocentric in light and purely egocentric in darkness. It is expected that the orientation system controlling behavior will also control PC mapping.

## Inv2

MENTAL REPRESENTATION IN THE BRAIN. *Chris Frith. Wellcome department of Cognitive Neurology, Institute of Neurology, Queen Square, London WC1N 3BG, UK*

For the neuroscientist studying the brain, neural activity in a certain region can represent an environmental feature, such as colour, or a behavioural response, such as movement in a certain direction. Such neural activity need not, however, represent anything to the animal in whose brain it occurs. A memory, on the other hand, arises from brain activity and represents a past event for the person to whom that event occurred. What can be said about the brain activity associated with these 'mental representations' which stand for something not currently present in the environment and about which we can give such detailed reports? Retrieval from episodic memory is reliably associated with activity in right prefrontal cortex. For example, right frontal activity is associated with successful recognition of a word and ERP measures suggest that this activity lasts well beyond the moment at which recognition occurs. We can speculate that the activity is associated with reconstruction of the context in which the word occurred enabling us to report details such as of the source of the word.

In the case of episodic memory, a mental representation enables us to reconstruct and reflect upon the details of a past event. Using much the same formulation we can also think of working memory and attention as depending upon mental representations: representations of the recent past or the expected future are held in mind and used to guide behaviour. Frontal activity is associated with both working memory and attention, but, in both, cases activity is seen in other areas as well. In studies of selective attention we are beginning to learn about the mechanisms by which frontal regions modulate activity in posterior regions where neural activity directly reflects features in the environment. It is these interactions between frontal and posterior regions which make mental representation possible.

## Inv3

DISTRIBUTED CORTICAL MEMORY. *J.M. Fuster. Brain Research Institute and Department of Posychiatry, School of Medicine, University of California, Los Angeles, CA 90024, USA*

Microelectrode recording in the monkey and imaging in the human indicate that the performance of memory tasks depend on the activation of widespread neuronal networks spanning posterior and frontal neocortex. The temporary retention of visual or tactile memorandum activates in sustained manner units in posterior and, additionally, in prefrontal cortex. The cooling of prefrontal regions affects the memory activity of posterior regions, and vice versa, posterior cooling affects the memory activity of prefrontal units. Concomitantly, the cortical cooling of circumscribed areas of either cortex, depending on the nature of the memorandum, reversibly depresses memory performance. In humans performing a visual-nonspatial-memory task, metabolic tomography shows the activation of widespread cortical regions that include dorsolateral prefrontal cortex. These data support the notion of widely distributed cortical memory networks. Such networks become temporarily activated in recall, working memory, and the formation of new memory. There is no empirical reason to postulate separate cortical networks for short- and long-term memory. Both may be served and supported by the same networks in different states of activation and consolidation.

## Inv4

NEURAL MECHANISMS FOR SYNTHESIZING SENSORY INFORMATION AND PRODUCING ADAPTIVE BEHAVIORS. *B.E. Stein. Department of Neurobiology and Anatomy, Bowman Gray School of Medicine/Wake Forest University, Winston-Salem, N.C. 27157-1010, USA*

The ability to integrate information from different sensory systems is a fundamental characteristic of the brain. Because different bits of information are derived from different sensory channels, their synthesis markedly enhances the detection and identification of external stimuli. The neural substrate for such "multisensory" integration is provided by neurons that receive convergent input from two or more sensory modalities. Many such multisensory neurons are found in the superior colliculus (SC), a midbrain structure that plays a significant role in overt attentive and orientation behaviors. The various principles governing the integration of visual, auditory, and somatosensory inputs in SC neurons have been explored in rat, cat, and rhesus monkey. Thus far, the evidence suggests a remarkable conservation of integrative features during vertebrate evolution. An example of one of the most robust of these principles is that: a striking enhancement in activity is induced in a multisensory neuron when two different sensory stimuli (e.g., visual and auditory) are in spatial concordance, whereas a profound response depression can be induced when these cues are spatially discordant. The most extensive physiological observations have been made in cat, and in this species the same principles which have been shown to govern multisensory integration at the level of the individual SC neuron have also been shown to govern overt attentive and orientation responses to multisensory stimuli. Most surprisingly, however, is the critical role played by association (i.e. anterior ectosylvian) cortex in facilitating these midbrain processes. In the absence of the modulating corticotectal influences, multisensory SC neurons in cat are unable to integrate the different sensory cues converging upon them, and overt multisensory attentive and orientation behaviors are compromised. This situation appears quite similar to that observed during early postnatal life. When multisensory SC neurons first appear, they are able to respond to multiple sensory inputs but are unable to synthesize these inputs to enhance or degrade their responses. During ontogeny, individual multisensory neurons develop this capacity abruptly, but at very different ages, until the mature population condition is reached after several postnatal months. It appears likely that the abrupt onset of this capacity in any individual SC neuron reflects the maturation of inputs from anterior ectosylvian cortex. Presumably, the functional coupling of cortex with an individual SC neuron is essential to initiate and maintain that neuron's capability for multisensory integration throughout its life.

Supported by NIH grants NS 22543 and EY 06562.

**Inv5**

LANGUAGE LEARNING IMPAIRMENTS: INTEGRATING BASIC SCIENCE, TECHNOLOGY AND REMIDIATION. *Paula Tallal. Center for Molecular and Behavioral Neuroscience, Rutgers University, Campus of Newark, University Heights, 197 University Avenue, Newark, New Jersey 07102, USA*

Timing cues present in the acoustic waveform of speech provide critical information for the recognition and segmentation of the ongoing speech signal. Research has demonstrated that deficient temporal perception rates, that have been shown to specifically disrupt acoustic processing of speech, are related to specific language-based learning impairments (LLI). Temporal processing deficits correlate highly with the phonological discrimination and processing deficits of these children.

Electrophysiological single cell mapping studies of sensory cortex in brains of primates have shown that neural circuitry can be remapped after specific, temporally cohesive training regimens, demonstrating the dynamic plasticity of the brain. Recently, we combined these two lines of research in a series of studies that addressed whether the temporal processing deficits seen in LLIs can be significantly modified through adaptive training aimed at reducing temporal integration thresholds.

Simultaneously, we developed a computer algorithm that expanded and enhanced the brief, rapidly changing acoustic segments within ongoing speech and used this to provide intensive speech and language training exercises to these children. Results to date from two independent laboratory experiments, as well as a large national clinical efficacy trial, demonstrate that dramatic improvements in temporal integration thresholds, together with speech and language comprehension abilities of LLI children, result from training with these new computer-based training procedures.

## Symposium 1

### Sym1/1

HIPPOCAMPAL INVOLVEMENT IN SPATIAL AND NON–SPATIAL RADIAL MAZE LEARNING IN INBRED MICE. *Wim E. Crusio[1] and Herbert Schwegler[2]. [1] Génétique, Neurogénétique et Comportement, CNRS UPR 9074, Centre de Transgénose, 3B rue de la Férrolerie, 45071 Orléans Cedex 02, France. [2] Institut für Anatomie, Leipziger Strasse 44, Universität Magdeburg, D-39120 Magdeburg, Germany*

Three-months old male mice from nine different inbred mouse strains were tested in two different spatial radial maze tasks: one in which the maze was turned by 45° between trials and one in which the maze was always placed in the same way. Only four out of eight arms contained food rewards, permitting simultaneous assessment of working (WM) and reference memory (RM) in both situations. Turning of the maze significantly decreased performance in a strain-dependent manner. Other animals from the same strains were processed histologically to estimate the strain-specific extents of the hippocampal intra- and infrapyramidal mossy fibre projections (IIPMF). The extents of the IIPMF correlated strongly with both WM and RM if the maze was turned between trials. Similar correlations were only found in the early phases of learning in the other condition. We conclude that the IIPMF are involved in spatial learning and that non-spatial within-maze cues may influence learning performance in some inbred strains.

| Factor: | I | II | III |
|---|---|---|---|
| Variable | | | |
| RM Non-spatial Changed | 0.95 | | |
| WM Non-spatial Changed | 0.93 | | |
| RM Non-spatial Not changed | 0.84 | | |
| WM Non-spatial Not changed | 0.79 | | |
| RM Spatial Turned Total | | 1.01 | |
| WM Spatial Turned Total | | 0.95 | |
| Spatial 8 arms | | 0.85 | |
| RM Not turned 1st week | | 0.83 | |
| WM Not turned 1st week | | 0.82 | |
| Unconfined 8 arm | | 0.81 | |
| IIP Mossy Fibres | | −0.92 | |
| RM Not turned 2nd week | | | 0.90 |
| WM Not turned 2nd week | | | 0.87 |
| Non-spatial 8 arms | | | 0.57 |

Harris-Kaiser orthoblique rotation; 3 factors with Eigenvalue >1; only loadings >0.40; variance explained: 95%; interfactor correlations <0.34.

These results were combined with those from previous studies, in which animals from the same inbred strains were tested in other, both spatial and non-spatial radial maze tasks: two spatial tasks in which all 8 arms of the maze were rewarded (one with, one without confinement to the central platform for 5 sec between subsequent arm choices); a non-spatial task in which all 8 arms were reinforced and in which animals had to push open doors in order to enter arms; 2 non-spatial tasks in which only 4 out of 8 arms were reinforced, black-and-white floor patterns identifying individual arms (one in which arms were interchanged between trials, one in which arms remained in the same configuration throughout).
A factor analysis rendered three factors: two representing non-spatial learning (factors I and III), one representing spatial learning (factor II). The IIPMF strongly loaded on the spatial factor only. We conclude that (1) spatial learning, in contrast to non-spatial learning, is a unitary process and (2) individual differences in the extent of the hippocampal IIPMF projection underlie individual differences in spatial learning abilities in the radial maze.

### Sym1/2

REVERSE GENETICS IN THE STUDY OF BRAIN AND BEHAVIOR: AN ALTERNATIVE APPROACH TO GENE TARGETING. *Robert Gerlai. GENENTECH, Inc., Neuroscience Department #72, 460 Point San Bruno Boulevard, South San Francisco, California 94080-4990, USA*

Gene targeting techniques using homologous recombination in embryonic stem (ES) cells allow the investigator to disrupt single genes in mammalian organisms including the mouse. These techniques offer an unprecedented precision with which one may manipulate genetic factors and investigate the in vivo effects of well defined mutations, an approach termed reverse genetics. Gene targeting has been suggested to be one of the most promising tools in the analysis of brain function and behavior. Molecular geneticists argued that the technique will obviate the lack of highly specific pharmacological tools to study various enzymes in the brain. Based on this promise, gene targeting has become very popular and widely used. However, as more gene targeting studies have been completed, it has become clear that this technique, as any other scientific method, is not faultless and has some disadvantages. One of the crucial problems with gene targeting is its temporal and spatial specificity. The mutation may appear in multiple brain regions or even in other organs and it may also be present from early developmental stages giving rise to complex, secondary phenotypical alterations due to compensatory mechanisms. This may be a disadvantage in the functional analysis of a large number of genes. For example, recently it has been discovered that proteins (e.g. neurotrophins, cell adhesion molecules and several receptor kinases) that were previously thought to have only a "developmental" role, may also be involved in neural and behavioral plasticity. Knocking out genes of such proteins may lead to major developmental alterations or even embryonic lethality in the mouse making it difficult to study the potential role such proteins may play in neural plasticity and behavior. Therefore, strategies alternative to gene targeting may need to be employed. I suggest one such strategy which is based on recombinant gene technology but combines it with the systemic delivery method typical of pharmacology studies. The approach is based on generation of a fusion protein, a chimera between the Fc portion of the human IgG molecule and a ligand binding domain of a receptor of interest. Such receptor-IgG fusion molecules have been shown to be stable in vivo and to exhibit high specificity and high affinity for the endogenous ligand of the receptor. However, they completely lack the ability to signal. Thus, if delivered into the brain, a fusion protein may be used to specifically block a signaling pathway of the receptor of interest. The fusion protein may be loaded into an osmotic minipump that offers a constant flow rate over an extended period of time (days or weeks). The minipump can be connected to a canula that may be lowered into the targeted brain region using a stereotaxic apparatus. Thus the timing and the location of delivery may be controlled in a precise manner. I will show that the above approach may be used successfully to block signaling between an Eph family receptor kinase and its ligand leading to a selective impairment in learning and synaptic plasticity in mice. The observed alterations provide the first conclusive pieces of evidence for the involvement of an Eph family receptor kinase in neural plasticity.

### Sym1/3

WHAT DO INBRED MICE ADD TO CURRENT INVESTIGATIONS ON THE NEURAL BASIS OF SPATIAL BEHAVIOURS ? *C. Rossi-Arnaud[1] & M. Ammassari-Teule[2]. [1] Department of Psychology, University of Rome "La Sapienza", Via dei Marsi, 78, 00185 – Rome, Italy. [2] Institute of Psychobiology and Psychopharmacology, C.N.R., Via Reno, 1, 00198 – Rome, Italy*

The large majority of experiments investigating the neural correlates of spatial behaviors in rodents has been conducted in animals belonging to oubred populations. This necessarily means that within-group individual levels of performance are distributed normally and that the behavioral profile of the group considered is given by the central value of the gaussian curve. Fundamental differences exist between these studies and those using inbred strains, that is, strains obtained by crossing male and female littermates presenting a same character to produce homozygous individuals. In that case, within-group individual differences are assumed to be minimal while between-group differences, that is, interstrain differences, reflect the variability evident in a natural outbred population. This non invasive method makes it possible to segregate high and low learner mice for particular tasks and search for the neural mechanisms subserving more specifically each performance level. The data presented here concern two strains of mice, C57BL/6 (C57) and DBA/2 (DBA), characterized by well-differentiated spatial abilities associated with strong neurobiological differences at the hippocampal level. Assuming that genetic differences in hippocampal functionality may lead to a different organization of brain circuits controlling in spatial performance, the main objective of this study was to compare the effect of selective lesions produced in areas connected to the hippocampus on radial maze performance in these two strains. Lesions were located in the frontal cortex, the posterior parietal cortex, the amygdala and the nucleus accumbens. The results showed that (1) both frontal cortex and amygdala lesions disrupted performance in the high performer C57 strain without producing any effect in the low performer DBA strain. Conversely, lesioning the posterior parietal area had a stronger impairing effect in DBA. Finally, lesioning the nucleus accumbens, paradoxically, improved DBA performance without affecting C57 performance. Taken together, these results indicate that comparing the effect of discrete lesions in inbred mice (1) reveals the existence of a large variability in the circuitry subserving each performance level which, generally, does not emerge from studies conducted in outbred animals (2) allows to identify which areas (parietal cortex) can gain control of performance and substitute a dysfunctioning hippocampus (3) points out the existence of brain sites (nucleus accumbens) susceptible to remove some inhibition and to improve spatial performance in poor performer mice. In particular, description of the specific circuitry subserving poor spatial abilities should be useful to approach the re-organization of brain systems in subjects presenting cognitive deficits.

## Sym1/4

GENETICS AND EPILEPSY: WHAT DO WE KNOW? *O. K. Steinlein. Institute of Human Genetics, University of Bonn, Wilhelmstr. 31, 53111 Bonn, Germany*

The term *idiopathic* is used to classify a major class of epilepsies which are not the consequence of recognisable diseases or brain damage, i.e. acquired disorders. It has long been recognised that genetic factors play a major role in the aetiology of idiopathic epilepsies. In the last few years rapid progress has been made towards the identification of epilepsy genes. However, the progress is impeded by the fact that most forms of idiopathic epilepsies are not due to single gene defects but show a complex pattern of inheritance. Especially the common forms, like juvenile myoclonic epilepsy or juvenile absence epilepsy, are probably due to the multiplicative contribution of several loci. Some of them might be responsible for a lowering of the seizure threshold, while others might influence the age of onset. One possibility to solve the problem of gene hunting in complex disorders is to first investigate those rare forms of the disease in which the phenotype seems to follow a simple mode of mendelian inheritance. Autosomal dominant nocturnal frontal lobe epilepsy (ADNFLE) and benign familial neonatal epilepsy (BFNC) are examples of such rare monogenic idiopathic epilepsies. Gene loci for BFNC and ADNFLE have been assigned to the same candidate region on chromosome 20q13.2–q13.3. Familial nocturnal frontal lobe epilepsy is the first idiopathic epilepsy for which a gene has been identified. A missense mutation (Ser248Phe) in the a4-subunit of the neuronal nicotinic acetylcholine receptor (CHRNA4) has been recently found in one large Australian pedigree (Steinlein et al. 1995 Nature Genetics, 11: 201–203). Expression studies showed that the mutant receptor, if compared with the wildtype receptor, has an accelerated desensitization rate as well as a markedly prolonged resensitization rate (Weiland et al., FEBS Letters, 1996, 398, 91–96). Thus, for the first time a relationship between an idiopathic epilepsy

and a specific gene has been found. To investigate the possible role of CHRNA4 in epilepsies with a more complex origin, an association study was performed in patients with idiopathic generalized epilepsies (IGE), mainly with juvenile myoclonic or childhood/juvenile absence epilepsy. Using a silent polymorphism in the coding region of the CHRNA4 gene a possible association was observed between the phenotype IGE and one allele of the polymorphism.

# Symposium 2

## Sym2/1

FROM VISION TO MOVEMENT: COMBINATORIAL COMPUTATIONS IN THE PARIETAL CORTEX. *R. Caminiti. Istituto di Fisiologia umana, Università "La Sapienza", Piazzale Aldo Moro 5 – 00185 Rome, Italy*

We move within a complex three-dimensional world anchored to objects which we identify and locate in space by using vision. Reaching movements are performed to bring the hand to objects of interest. It is believed that the distributed cortical network underlying visual reaching transforms the information concerning the spatial location of the target into an appropriate motor command. This problem can be decomposed into two related questions, the first of which concerns the *substrata* whereby visual information reaches the motor apparatus. The second relates to the *form* of this information, once it is available to frontal motor areas, that is, to the transformation from visual to motor coordinates believed to occur at the interfaces between vision and movement. Both of these questions remain for the most part unanswered. The set of cortical areas and pathways by which the information on target location is relayed from the visual areas of the occipital lobe to the motor areas of the frontal lobe have been identified only recently.

Different types of information must be combined within this parietofrontal network in order to compute appropriate motor commands to move both eye and hand to visual targets, that is, the visually-derived information about target location and the somatic information about the position of the arm in space. This combination occurs at a very early stage of the information processing flow, in the cortex of the rostral bank of the parieto-occipital sulcus and in area 7 m, in the medial wall of the superior parietal lobule. We have used different behavioral tasks to dissociate eye from hand contributions to cell activity in these areas. In the parieto-occipital cortex, once believed to be visual and oculomotor in nature, we have found reaching-related activity which results from a combination of gaze and hand signals. Area 7 m, which is less heavily linked to the visual world than V6 and V6A, is characterized by populations of neurons related to eye position and movement, to hand position and movement and to a combination of visuomanual and oculomotor information.

These observations suggest that the computation from visual input to motor output occurs through a process which uses, probably through re-entrant signals, kinesthetic and motor information at a very early stages of composition of the motor command. They also assign a novel role to parieto-occipital and mesial parietal cortices.

## Sym2/2

NEURAL MECHANISMS OF SPACE REPRESENTATION IN THE VENTRAL INTRAPARIETAL AREA OF THE MONKEY. *J.-R. Duhamel[1] and F. Bremmer[2]. [1] LPPA – College de FRANCE, 11 place Marcelin Berthelot, F-75005 Paris France; [2] Ruhr-University Bochum, Dept.Zoology & Neurobiology, D-44780 Bochum, Germany*

Visually-guided behavior requires a transformation from a retinocentric frame of reference into a frame which is appropriate for movement of the selected body part, be it the eye, the head or the trunk. It is often assumed that this is achieved through a series of intermediate computations, such as the a calculation of the object's head-centered coordinates. One view of how this might be accomplished holds that higher-order spatial representations emerge from the distributed activity of populations of neurons which combine information from retinal and extra-retinal sources. In support of this concept, modeling experiments have shown that neural networks trained to encode spatial locations in a head-centered frame of reference generate units whose properties resemble a particular class of neurons commonly found in the monkey parietal cortex: visual neurons whose receptive field (RF) is retinocentric but whose activity level is gated by eye position. An alternative, though not completely incompatible, view is that in the occipito-parietal visual pathways, multiple specialized areas process sensory information in connection with specific motor behaviors, and each area represents visual space using a encoding formats which can be expressed in the response properties of individual neurons. The lateral intraparietal area (LIP), which is involved in saccadic eye movements, contains neurons encoding visual space in eye-centered coordinates, and those coordinates are updated in conjunction with intended eye movements. In contrast the adjacent ventral intraparietal area (VIP) is organized quite differently. This area is involved in a multisensory analysis of movement. Consistent with a strong connection with area MT, most VIP neurons are sensitive to the direction and speed of moving objects. Although the prevalent sensory modality is visual, VIP neurons often respond equivalently to visual, tactile, and vestibular stimulation, with parallel preferred directions for all three modalities. The somatic representation emphasizes the face region, and neurons with visual and tactile responses have spatially-congruent receptive fields, i.e. cells with foveal visual RFs have tactile RFs located on the mouth, and cells with peripheral visual RFs have somatic RFs located on the side of the head. In addition, using a quantitative mapping procedure, we found that area VIP contains visual neurons whose RF is spatially-invariant and can represent the azimuth and/or the elevation of visual stimuli in head-centered coordinates. Finally, in both LIP and VIP, neurons display visual responses whose intensity depends on orbital eye position. Thus although both areas have the neural building blocks for generating potentially a population-level description of space, VIP stands out as also being able to directly encode spatial locations using single neurons. Further work is needed to clarify the role of eye position and other extra-retinal signals recently described in several cortical areas, as well as the relations between explicit (single cell level) and implicit (neural population level) frames of reference for space representation.

## Sym2/3
A PARIETO-PONTO-CEREBELLAR PATHWAY FOR GOAL-DIRECTED EYE MOVEMENTS. *P. Thier[1], P. W. Dicke[1], C. Schwarz[1], S. Barash[2]. [1] Sektion für Visuelle Sensomotorik, Neurologische Universitätsklinik, 72076 Tübingen, Germany. [2] The Weizmann Institute, 76100 Rehovot, Israel*

The primate dorsolateral pontine nucleus (DLPN) is commonly considered to be a key link in a cerebro-ponto-cerebellar pathway for smooth pursuit eye movements, a pathway assumed to be anatomically segregated from the tegmental circuits subserving saccades. However, the existence of afferents from several cerebrocortical and subcortical centers for saccades including parietal area LIP and the frontal eye fields suggests that the DLPN might contribute to saccades as well. In order to test this hypothesis, we recorded from the DLPN of monkeys trained to perform smooth pursuit eye movements as well as visually and memory-guided saccades. Out of 167 neurons so far isolated from the DLPN and the neigboring regions of the dorsal pontine nuclei, 66 were responsive in oculomotor tasks. 32 were exclusively activated by saccades, 18 exclusively by smooth pursuit and 16 neurons showed sensitivity to both saccades and smooth pursuit. When tested in the memory saccade paradigm, 16 out of 48 of the saccade-related neurons showed significant activity in the memory period. Our finding of saccade-related activity at the level of the DLPN in combination with the existence of strong anatomical input from saccade-related cerebrocortical areas such as parietal area LIP suggests that the DLPN serves as a precerebellar relay for both pursuit and saccade-related information originating from cerebral cortex, bypassing the classical tecto-tegmental circuitry for saccades. What is the functional role of the DLPN in this pathway? One possible answer is that the PN might contribute to the shaping of cerebellar response patterns by integrating signals derived from anatomically segregated parts of cerebral cortex. This conclusion is suggested by the aforementioned finding of a substantial number of single neurons in the DLPN responding to saccades as well as to smooth-pursuit. Such neurons may be useful to coordinate the contributions of saccades and smooth-pursuit to the tracking of slowly moving targets, which usually consists of segments of smooth-pursuit interrupted by catch-up saccades. Sensitivity to both saccades and pursuit is a property, DLPN neurons share with many Purkinje-cells in their major projection zone in vermal lobuli VI and VII. On the other hand, we are not aware of any descriptions of cerebrocortical or collicular neurons, responding to both saccades and smooth pursuit. Rather, cerebral cortex exhibits an organization which is characterized by multiple *eye fields*, devoted either to saccades or to smooth pursuit. The expression of combined sensitivities to pursuit and saccades at the level of single units in the DLPN therefore requires integration of information derived from segregated cerebrocortical *eye fields*. Our own anatomical work suggests that afferents originating from different parts of cerebral cortex do not converge on single pontine nuclei neurons. Hence, integration of signals having different cerebrocortical sources requires exchange of signals between pontine nuclei neurons either by intrinsic circuitry or by a feedback loop through the cerebellum.

Supported by the Deutsche Forschungsgemeinschaft and the German Israeli Foundation.

## Sym2/4
SOMATOSENSORY AND MOTOR PROCESSES IN SPATIAL HEMINEGLECT. *Giuseppe Vallar. (Dipartimento di Psicologia, Universita' di Roma "La Sapienza", Via dei Marsi 78, 00185 ROMA, and IRCCS Clinica S. Lucia, Roma, Italy)*

In patients with lesions in the right hemisphere, frequently involving the posterior parietal regions, left-sided somatosensory, visual and motor deficits reflect not only a disorder of primary sensory processes, but have also a higher order component related to a defective spatial representation of the body. This additional factor, related to right brain damage, is clinically relevant: contralesional hemianaesthesia, hemianopia and hemiplegia are more frequent in right brain-damaged patients, than in patients with damage to the left side of the brain. Three main lines of investigation suggest the existence of this higher order pathological factor. (i) Right brain-damaged patients with left hemineglect may show physiological evidence of preserved processing of somatosensory stimuli, of which they are not perceptually aware. Similar results have been obtained in the visual domain. (ii) Direction-specific vestibular, visual optokinetic and somatosensory/proprioceptive stimulations may displace spatial frames of reference in right brain-damaged patients with left hemineglect, reducing or increasing the extent of the patients' ipsilesional rightward directional error, and bring about similar directional effects in normal subjects. These stimulations, which may improve or worsen a number of manifestations of the neglect syndrome, (e.g., extra-personal and personal hemineglect) have similar effects on the severity of left somatosensory deficits (defective detection of tactile stimuli, position sense disorders). Visuo-spatial hemineglect and the somatosensory deficits improved by these stimulations are independent, though related, disorders, however. (iii) The severity of left somatosensory deficits is affected also by the spatial position of body segments. A general implication of these observations is that spatial (non-somatotopic) levels of representation contribute to corporeal awareness. The neural basis of these spatial frames includes the posterior parietal and the pre-motor frontal regions. These spatial representations could provide perceptual-pre-motor interfaces for the organisation of movements (e.g., pointing, locomotion) directed towards targets in personal and extra-personal space. In line with this view, recent evidence indicates that the sensory stimulations which modulate left somatosensory deficits may affect left motor deficits in a similar, direction-specific, fashion.

## Symposium 3

### Sym3/1

NEUROIMAGING STUDIES OF COLOR PERCEPTION, SELECTIVE ATTENTION TO COLOR, AND COLOR KNOWLEDGE. *J.V. Haxby. Section on Functional Brain Imaging, Laboratory of Brain and Cognition, NIMH, Bldg 10, Rm 4C104, Bethesda, MD 20892-1366, USA*

The representation of color in the human brain can be studied noninvasively with functional neuroimaging. Two functional neuroimaging methods, positron emission tomography (PET) and functional magnetic resonance imaging (fMRI) have been used to measure local hemodynamic changes associated with color processing. These hemodynamic changes are used as indices of changes in neural activity integrated over time in a volume of tissue. PET studies have demonstrated that both passive viewing of color and selective attention to color activate an extrastriate region in ventral medial occipital cortex (Zeki et al. *J Neurosci*, 1991; 11:641–649; Corbetta et al. *J Neurosci*, 1991; 11:2383-2402). A fMRI study of selective attention to color has corroborated this finding and identified the anatomical location of this region with greater precision (Clark et al. *Hum Brain Mapp*, in press). In this study, subjects performed a delayed match-to-sample test of selective attention to the color or the identity of color-washed faces. Selective attention to color consistently activated a region in the collateral sulcus more than did selective attention to face identity. The representation of semantic knowledge about the colors of objects was investigated with PET by asking subjects to generate the name of a color typically associated with different objects, presented as black and white line drawings or as words (Martin et al. *Science*, 1995; 270:102–105). A region in the fusiform gyrus was more activated by this condition than by generating the name of the object or by generating an action associated with the object. This region was approximately 3 cm anterior to the collateral sulcal area activated by color perception. These studies show that the representation of perceived color and the representation of knowledge about color are associated with activity in specific extrastriate regions in the ventral visual pathway. Selective attention to the color of visual stimuli can modulate the strength of activity in the area associated with color perception. Knowledge that has been acquired from experience with color is represented in nearby cortex, and the strength of activity associated with this representation can also be modulated by selective attention to the color attributes of a semantic memory.

### Sym3/2

COLOUR, FORM AND MOTION IN CEREBRAL ACHROMATOPSIA. *C.A. Heywood[1], R.W. Kentridge, and A. Cowey[2]. [1] Dept. of Psychology, Science Laboratories, South Road, Durham, DH1 3LE. [2] Dept. of Experimental Psychology, South Parks Road, Oxford, OX1 3DU*

Patients with cerebral achromatopsia, resulting from damage to ventromedial occipital cortex, are quite unable to chromatically order, or discriminate, hue. Their perceptual world is colourless. Nevertheless, they show a strikingly intact ability to use wavelength information to detect stimuli defined by isoluminant chromatic changes to the extent that their chromatic contrast sensitivity can be indistinguishable from that of the normal observer. Moreover, discrimination of the direction of motion of a chromatic grating, under circumstances that require knowledge of the sign of the colours (what is red, and what is green) of which it is composed, is also intact.

A possible contributor to the detectability of chromatic gratings is the subadditive nature of colour combination such that mixtures of red and green look perceptually dimmer than that expected from the simple addition of luminances. This subadditivity is believed to reflect colour-opponent interactions between the outputs of long- and medium-wavelength cones, respectively. We performed a first order compensation for such subadditivity in chromatic gratings and demonstrated that their detection was not abolished in an achromatosic patient. In addition, we used a 2-alternative forced-choice

procedure with two achromatopsic patients, who were required to judge the apparent relative velocity of two drifting gratings with different degrees of compensation. It is well- known that isoluminant gratings, constructed by adding a red and green sinusoidal grating of identical peak luninaces in antiphase, appear to drift substantially slower than an achromatic grating with the same velocity. Adding 2f luminance compensation to an isoluminant grating of spatial frequency, f, resulted in an identical minimum of perceived velocity at a compensation contrast of 5% in both achromatopsics and normal observers. Furthermore, while compensation for subadditivity did not compromise grating detection at low contrasts, such correction severely affected motion detection.

We conclude first that subadditivity, which results from colour-opponent P-channel processes, can influence motion judgements conventionally assigned to the M-channel. Second, detection of sinusoidally modulated chromatic gratings in achromatopsic patients is not as a result of subadditivity. Finally, the ability to extract motion from chromatic differences alone is intact in achromatopsia, in a manner which is comparable to that of the normal observer. Finally, and more generally, wavelength processing continues to contribute to several aspects of visual processing even when colour is not perceived.

### Sym3/3

COLOUR CONSTANCY: WHY, HOW AND WHERE? *A.C. Hurlbert*, D. I. Bramwell*, Alberto Palomares**, Jose Artigas**. * Physiological Sciences, University of Newcastle-upon-Tyne, England NE2 4HH. ** Departamento de Optica, Universitat de Valencia, 46100-Burjassot, Spain*

Objectives. Is colour constancy the consequence of a neural mechanism that evolved specifically to achieve it, or the by-product of another design? Theoretical studies suggest that the cone-opponent transformations required for optimisation of information transmission may also achieve colour constancy, by producing channels in which the colour descriptors remain largely invariant under natural illuminant shifts. Such transformations occur at an early stage of visual processing. Other theoretical studies and some experimental results – particularly those implicating long-range spatial interactions or recognisable scene features such as specular highlights or mutual illumination – suggest that higher-level cortical mechanisms may contribute to colour constancy. We performed a series of asymmetric matching experiments of surface colours against simple and complex backgrounds and evaluated several models of early cone-opponent and non-opponent processing for predicting the data. Methods. Test surfaces were either 2-deg square simulated Munsell papers or depictions of real objects with Munsell paper surface reflectances appropriate to their real-world colours. The surfaces appeared against uniform, multi coloured checkerboard (Mondrian) or complex real backgrounds, all under simulated daylight illuminants, displayed on a CRT screen. Observers viewed the test surface/background under the test illuminant with the left eye, and the standard surface/background under the standard illuminant with the right eye in a fused haploscopic display. To stabilise adaptation, we held test and standard illuminants fixed throughout each experimental session. Observers adjusted the colour of the standard surface to match that of the test surface. A subset of the observers' matches for a range of surface colour/illuminant combinations were used to train an artificial neural network. The transformation model represented by the network, as well as several analytically derived transformation models, were then used to predict observers' matches. Results. For both uniform and Mondrian backgrounds, observers' matches were predicted qualitatively by the simple ratio of non-opponent cone excitations for the surface and background. The predictions improved for two-stage models and neural networks that incorporated cone-opponent ratios between surface and background. Observer matches for real objects in complex scenes were not well predicted by either non-opponent or cone-opponent ratios. Conclusions. Colour constancy for simple artificial scenes may be entirely accounted for by low-level mechanisms that compute ratios between surface and background in cone-opponent channels. These mechanisms are not sufficient to account for the preservation (or lack thereof) of colour appearance of real surfaces in complex scenes that include shadows, highlights, recognisable objects, and other factors that may influence scene interpretation.

**Sym3/4**
COLOUR MEMORY AND COLOUR CONSTANCY DEFICITS IN HUMANS AND MACAQUE MONKEYS FOLLOWING DAMAGE TO VISUAL AREA V4. *V. Walsh[1] & S. Clarke[2]. [1] Dept of Experimental Psychology, University of Oxford, South Parks Rd., Oxford OX1 3UD, U.K. [2] Division de Neuropsychologie, CHUV, 1011 Lausanne, Switzerland*

Human subjects who had suffered damage to regions of the lingual and fusiform gyri were tested on a range of tests of colour processing, including discrimination, naming, memory and constancy. We were concerned to identify which cortical loci were necessary for these different aspects of colour perception and to compare the performance of the neuropsychological patients with that of monkeys who had suffered lesions of areas V4 or TEO. Three patients suffered colour constancy deficits and all three of these were deficient in colour memory. Two of these patients had suffered bilateral damage to the posterior lingual and fusiform gyri and one to the same region in the right hemisphere. A similar lesion restricted to the left hemisphere produced a deficit in colour naming and colour memory but not in colour constancy. Macaque monkeys with bilateral lesions of visual area V4 also had deficits in colour memory and colour constancy and TEO lesions produced deficits in colour memory. In addition to the similarities of the deficits observed in human and macaque subjects an important negative similarity is that none of the subjects were achromatopsic. We suggest that V4 in man and monkey is important for colour memory and colour constancy.

## Symposium 4

**Sym4/1**
NEUROIMAGING STUDIES OF LONG TERM MEMORY. *P.C. Fletcher. Wellcome Department of Cognitive Neurology, Institute of Neurology, Queen Square, London, UK, WC1N 3BG*

Neuroimaging studies of long term memory have emphasised the role of prefrontal cortex (PFC). An attraction of such studies has been their ability to dissociate brain systems associated with the encoding stage from those associated with the retrieval stage. An unpredicted, but consistent, finding has been that of a lateralisation of PFC function with left PFC activation being predominant at encoding and right PFC activation at retrieval. The functional significance of this phenomenon has yet to be clarified. We suggest that the lateralisation of frontal function reflects differing executive processes which are necessary for optimal encoding and retrieval. The former stage emphasises, among other things, the organisation of study material while the latter may include a monitored search of the contents of memory. In studies of encoding and retrieval of verbal material, we have manipulated the requirement to use such processes and provide evidence that a task which emphasises the need to set up an organisational structure for study material is associated with a significantly greater degree of left PFC activation. Further, a distracting task, which interferes with such a task, produces an attenuation of this activity. At the retrieval stage, the need to set up a monitored search is associated with right PFC activation. This activation appears to be specific to the dorsolateral region whereas a retrieval task which minimises the monitoring needs produces a ventrolateral activation.
Although preliminary, our results, and those of other groups, suggest that functional neuroimaging techniques have an important role to play in clarifying the roles of PFC in long term memory function.

**Sym4/2**
ANATOMICAL MODELS OF VERBAL WORKING MEMORY: NORMAL AND PATHOLOGICAL. *E. Paulesu. INB-CNR, Istituto Scientifico H San Raffaele, via Olgettina 60, 20132 Milano, Italia*

In recent years attempts have been made to map working memory processes onto normal and dysfunctional brain anatomy. I shall review the available evidence of functional imaging experiments on the phonological loop, the verbal slave system of working memory. Psychological and neuropsychological investigations propose that the phonological loop comprises a rehearsal system and a short-term phonological store 1, 2 . Review of the functional imaging experiments leads to the following suggestions.
i) The data support a multicomponent model for the phonological loop, in which the subvocal rehearsal system and the phonological store are anatomically distinct. In normal subjects, subvocal rehearsal and short-term phonological storage rely on left ventral premotor cortex and left temporoparietal cortex respectively 3, 4.
ii) As predicted by cognitive psychology and neuropsychology, the phonological code(s) of the phonological loop are independent from early acoustic codes and primary auditory areas 5.
iii) The functional/anatomical segregation of the two main components of the phonological loop is also confirmed by studies of developmental dyslexia 6 and patients with acquired brain damage and pathological verbal span7.
I will also discuss to what extent the anatomical model of the phonological loop has explanatory value to describe the physiological bases of the acquisition of new vocabulary 8.

### References

1. Baddeley, A.D. Working memory . Oxford University Press, Oxford (1986).
2. Shallice, T. & Vallar, G. in Neuropsychological impairments of short-term memory (eds. Vallar, G. & Shallice, T.) Cambridge University Press, New York (1990).
3. Paulesu, E., Frith, C. & Frackowiak, R. Nature 362, 342–344 (1993).
4. Démonet, J., Fiez, J., Paulesu, E., Petersen, S. & Zatorre, R. Brain and Language 55, 352-379 (1996).
5. Paulesu, E., Bottini, G. & Frackowiak, R. in Contemporary Behavioral Neurology (eds. Trimble, M. & Cummings, J.) 49–89, Butterworth-Heinemann, Boston (1996).
6. Paulesu, E., et al. Brain 119, 143–157 (1996).
7. Paulesu, E., Shallice, T., Frackowiak, R.S., Frith, C.D. (in preparation).
8. Paulesu, E., Vallar, G., Signorini, M., Burani, C., Buechel, C., Perani, D., Fazio, F. (in preparation).

**Sym4/3**
ELECTROPHYSIOLOGICAL STUDIES OF EPISODIC MEMORY. *Michael D. Rugg. Wellcome Brain Research Group, School of Psychology, University of St Andrews, St Andrews KY16 9JU, Scotland*

Event-related brain potentials (ERPs) are an attractive means of studying human brain activity elicited during memory tasks because of their non-invasiveness, their excellent temporal resolution, and the ease with they can be obtained for items associated with different behavioural outcomes (e.g. 'hits' vs. 'misses' in recognition memory). Furthermore, as knowledge about the intracerebral generators of memory-related ERP effects increases, ERPs will offer a means to study, in 'real-time', the time course and interaction of specific neural circuits supporting memory. This presentation will focus on findings from recent ERP studies of episodic memory retrieval.
In a number of studies, ERPs elicited by correctly recognised items have been formed separately according to whether or not recognition was accompanied by the retrieval of the prior episode involving the item ('recollection'). This segregation was achieved by modifying

the test phase of recognition memory tasks so that subjects first make a standard 'old/new' judgement, and then, for items judged old, a second judgement requiring knowledge about the context in which the item was experienced. Compared to items correctly judged new, ERPs to old items show several characteristic differences. These differences are larger, and in some cases are only present, in waveforms evoked by items for which information about the learning context can be recovered. The results of these studies give insights into the time-course and possible neural underpinnings of retrieval from episodic memory, and have in addition helped shed light on the causes of the decline in episodic memory function that accompanies normal ageing. These results will be reviewed, and their relationship to findings from studies employing functional neuroimaging methods discussed.

## Sym4/4
FUNCTIONAL MAGNETIC RESONANCE IMAGING OF THE HUMAN BRAIN: DATA ACQUISITION AND ANALYSIS. *Robert Turner. The Wellcome Department of Cognitive Neurology, Institute of Neurology, Queen Square, London WC1N 3BG, UK*

Techniques of functional brain mapping using magnetic resonance imaging (fMRI) are described. Gradient-echo MRI images are sensitive to local changes in magnetic susceptibility, which can occur when the oxygenation state of blood changes. Cortical neural activity causes increases in blood flow, which usually result in changes in blood oxygenation. Hence changes of image intensity can be observed, giving rise to the so-called Blood Oxygenation Level Dependent (BOLD) contrast technique. Use of echo-planar imaging methods (EPI) allows the monitoring over the entire brain of such changes in real time. A temporal resolution of 1–3 seconds, and a spatial resolution of 2 mm in-plane, can thus be obtained. Generally in a brain mapping experiment hundreds of brain image volumes are acquired at repeat times of 1–6 seconds, while brain tasks are performed. In the analysis of such data it is often useful to create statistical maps of image differences, whether of intensity or morphology. The technique known as Statistical Parametric Mapping (SPM), based on robust multilinear regression techniques, has become the method of reference for analysis of positron emission tomography (PET) image data. The special characteristics of fMRI data require some modification of SPM algorithms and strategies, and the MRI data must be gaussianized in time and space to conform to the assumptions of the statistics of Gaussian random fields. The steps of analysis comprise: removal of head movement effects, spatial smoothing, and statistical inference, which includes temporal smoothing and removal by fitting of temporal variations slower than the experimental paradigm. By these means, activation maps can be generated with great flexibility and statistical power, giving probability estimates for activated brain regions based on intensity or spatial extent, or both combined. Recent studies have shown that patterns of activation obtained in human brain for a given stimulus are independent of the order and spatial orientation with which MRI images are acquired, and hence that inflow effects are not important for EPI data with a TR much longer than T1.

## Sym4/5
IMAGING RECOVERY FROM STROKE. *Cornelius Weiller. Neurologische Klinik, Friedrich-Schiller-University, Jena, Deutschland*

Recovery of lost function through a persistent structural lesion in the central nervous system is accompanied by a complex pattern of changes in the organisation of the brain. Changes depend on the site of the lesion, are individually different and found in both hemispheres, the damaged and the sound one. The main theme is re-weighting of activity between the various representational levels within a preexisting, widespread and bilateral organised and parallel processing network. This implies changes in rest-activity wth increased or decreased blood flow and altered activation patterns during performance of the restituted function.
Within the primary motor system an activation at the rim of the infarct, extension into neighbouring representations, which outflow is not disturbed, altered recruitment pattern of motor cortex neurons, and recuitment of ipsilateral direct descending corticospinal tract pathways originating in the sound hemisphere are found.
Disruption of the primary system leads to increased activity in secondary of higher order areas and activation of areas related to attentional and intentional mechanisms.
Language is represented in a bilateral network including Wernicke's, Broca's area, left DLPFC and their homologues in the right hemisphere and additional areas in the parietal and temporal lobe. A verb generation paradigm in recovered patients with classical Wernicke s or Broca s aphasia and corresponding lesions showed increased activity in the remaining parts of the network in the left and the right hemisphere alike.
Proprioreceptive stimulation induced by passive movement can be used to assess reorganisation of the brain during the very early stage of rehabilitation (1 week). During complete plegia about 50% of patients showed signs of early reorganisation, which correlate with recovery after 3 months.
Short term (10×15 minutes) intense language comprehension training during PET scanning induced restitution of impaired comprehension in patients with Wernicke type apasia. This improvement correlated with right hemisphere activation foci only.
Future harvest may allow the pathophysiological evaluation of therapeutic measures, assessment of individual prognosis and allocation of patients to their optimal therapy.

## Symposium 5

## Sym5/1
MOTORIC REPRESENTATION OF OBJECTS. *Giacomo Rizzolatti. Istituto di Fisiologia Umana, Università di Parma, Via Gramsci 14, I-43100 Parma, Italy*

We know much more about grasping mechanisms than about reaching mechanisms. Although this might seem to be surprising, in fact is not. Grasping implies a transformation of an object (a real thing) into a movement, while reaching implies a transformation of space (an abstract construct) into a movement. Basing the theory on grasping data, it emerges that goal-directed actions are organized using neurons each of which codes a discrete motor act. These discrete acts are stored in the premotor cortex. The best example is monkey area F5 which stores hand movements (motor vocabulary).
In my presentation I will discuss first how F5 motor vocabulary can be visually addressed. I will show that a set of F5 neurons respond to visual presentation of objects ("visuomotor" neurons). The response are present when the object presentation is followed by grasping as well as during object fixation. Most "visuomotor" neurons discharge selectively in response to one object or to a small set of objects. Neurons that code a given grip discharge selectively at the presentation of objects that are grasped using that grip. On the basis of these and other findings, I will submit that F5 neurons contribute to object categorization by matching coded movements with the description of object intrinsic properties.
In the second part of my talk I will present two experiments carried out in humans showing: a) that visual presentation of an oriented geometric figure primes automatically a motor action congruent with figure orientation and b) that the preparation of a specific grasping movement influences the speed of analysis of a figure pictorially congruent with the object to be grasped. I will conclude by discussing how the motoric description of objects may intervene in the organization of concept development.

## Sym5/2
THE ROLE OF PERIETAL CORTEX: COMPARATIVE EVIDENCE FROM LESION AND ACTIVATION STUDIES. *H.J. Freund. Neurologische Klinik der medizinischen Einrichtungen der Universität Düsseldorf, Düsseldorf, Germany*

Activation of the human parietal lobe in conjunction with premotor and primary motor cortex is seen in a variety of sensorimotor tasks. The activations show modality specific patterns for somatomotor and visuomotor behaviour. For reaching and grasping they cluster bilaterally in a more posterior and anterior zone of intraparietal sulcus. Complex bimanual visuomotor tasks activate more medial parietal and frontal medial wall areas. The adjustment of visual and proprioceptive information required for the remapping of visuomotor coordinate transformation activated superior posterior parietal cortex. This modular activation pattern is complemented by data from patients with small unilateral lesions of these areas showing impairment of the respective functions on the contralesional side. In contrast, the effects of lesions of the inferior parietal lobule (IPL) are side dependent. On the right side IPL damage preferentially affects spatial functions, whereas left sided IPL damage compromises motor behaviour at the most global level: These patients with ideational apraxia are unable to organise their motor behaviour in a purposive manner even for everyday activations like dressing. This emphasises the role of the left IPL for the elaboration of motor concepts. The combined activation and lesion approach supports the concept that different types of sensorimotor behaviour are spatially encoded by the interaction of distinct functional modules in parietal and premotor areas before they are relayed to topographically organized motor representations, and that some of these functions are lateralized.

## Symposium 6

### Sym6/1
MAJOR BIOLOGICAL ACTIONS OF CCK. *H. Fink, A. Rex, M. Voits, J.-P. Voigt. Institute of Pharmacology and Toxicology, Medical Faculty Charité, Humboldt University at Berlin, D-10098 Berlin*

Cholecystokinin (CCK) is one of the first discovered gastrointestinal hormones and it is one of the most abundant neuropeptides in the brain.
Two types of CCK receptors have been identified: CCK-A receptors are mainly located in the gastrointestinal system but also found in some areas of the brain. CCK-B receptors are widely distributed in the brain. Major biological actions of CCK are the reduction of food intake and the induction of anxiety related behaviour. Inhibition of feeding is mainly mediated by the A-type receptors, whereas anxiety-like behaviour is induced by stimulating B-type receptors.
New findings are presented on the actions of the biologically active CCK fragments, CCK-8S, CCK-4 and A 71378.
It is demonstrated that the hypophagic effects of CCK are strongly dependent of the experimental design, sex and age of the rats.
Food intake measured during night or after food deprivation is reduced by CCK-8S in young adult and aged rats, whereas under fixed feeding conditions CCK-8S does not inhibit food intake in the young adult rats.
The sensitivity to the hypophagic CCK effect increases with age in male and female rats, however female rats are less sensitive to the CCK action. Furtheron, using a nongenetic and less stressful model of obesity due to unspecific postnatal overfeeding, the satiating effect of moderate CCK-8S doses is weaker in obese than in normal rats. Again, the hypophagic effect is more pronounced in male than in female obese and normal rats.
Considering that the aversive reaction in rats is markedly influenced by strain and breeding line variations research findings in this area are critically reviewed. It is shown that anxiety-like symptoms can only be induced by selectively acting CCK-B agonists, whereas mixed CCK-A and -B agonists and selective CCK-A agonists fail to change the behaviour in anxiety tests.

CCK-4 induces stable and reproducible anxiety-related behaviour only in certain rat strains, independent from the basal level of anxiety and the responsiveness to other effective drugs. Moreover, CCK-4 effects can be demonstrated in a modified open-field, in the ultrasonic vocalisation test in rat pups, on the elevated plus maze and in the black and white box, but not in the social interaction test.
The literature on the behavioural pharmacology of CCK is rife with inconsistency and contradiction. In principle, the major biologically actions of CCK depend on the receptor selectivity of the used CCK fragments, on organismic and procedural variables, influencing behaviour and CCK response in the tests.
In CCK research more attention has to be paid to the importance of these methodological factors.

### Sym6/2
MELANOCORTINS IN BRAIN AND BEHAVIOR. *W.H. Gispen and R.A.H. Adan. Rudolf Magnus Institute for Neurosciences, Medical Pharmacology, Utrecht University, Universiteitsweg 100, 3584 CG Utrecht, The Netherlands*

Peptides related to ACTH and MSH are long known to exert a variety of effects on the brain and modulate nervous behaviors in animal and men. In recent years five melanocortin (MC) receptors have been cloned. In the brain the $MC_3$ and $MC_4$ receptor subtypes are expressed. Receptor activation studies in vitro lead to the identification of development of more or less specific agonists and antagonists. Using these tools it could be demonstrated that the $MC_4$ receptor mediates novelty as well as MSH-induced excessive grooming in rats. However, this relationship is not unique as the brain $MC_4$ receptor is also implicated in body weight control and neurotrophic activation. Following intracerebroventricular administration, a-MSH exerts strong anti-pyretic effects. One of the first behavioral activities of ACTH/MSH-like peptides reported was that on the acquisition and extinction of avoidance behavior. The peptide structure activity relationship of these effects suggest that in addition to $MC_3$ and $MC_4$ in the brain there must be as yet an unknown MC receptor that recognizes the ACTH/MSH$_{4-9}$ analog Org 2766.

### Sym6/3
ROLE OF DYNORPHIN AND ENKEPHALIN IN THE REGULATION OF STRIATAL OUTPUT PATHWAYS AND BEHAVIOR. *H. Steiner[1], C. R. Gerfen[2]. [1] Department of Anatomy and Neurobiology, University of Tennessee, College of Medicine, Memphis, TN 38163, USA. [2] Laboratory of Neurophysiology, National Institute of Mental Health, Bethesda, MD 20892, USA*

The two subtypes of striatal projection neurons contain opioid peptides in addition to the neurotransmitter GABA. Neurons that send axons to the entopeduncular nucleus and/or the substantia nigra (striatonigral neurons) express dynorphin, whereas neurons that project to the globus pallidus (striatopallidal neurons) contain enkephalin. In a series of studies, we have investigated the role of these opioid peptides, using induction of immediate-early genes (IEGs) such as c-*fos* and *zif 268* as a functional marker. Gene expression was measured with quantitative in situ hybridization histochemistry. Our results show: (1) The magnitude of IEG induction by the psychostimulant cocaine, which occurs in striatonigral neurons and is mediated by D1 dopamine receptors, is inversely related to the level of dynorphin expression in a given striatal region. (2) Repeated cocaine treatment that produces behavioral sensitization results in increased dynorphin expression in striatonigral neurons and suppressed IEG inducibility. (3) This suppression of IEG induction can be mimicked with intrastriatal administration of a dynorphin (kappa opioid receptor) agonist prior to cocaine injection, demonstrating that kappa receptors located in the striatum are involved. Also, unilateral intrastriatal infusion of the dynorphin agonist induced ipsiversive turning in animals injected with systemic cocaine. (4) Experiments with selective D1 receptor agonists in animals without dopamine terminals (after a 6-OHDA lesion) indicate that dynorphin/kappa agonists interact with the dopamine transmission presynaptically (inhibition of dopamine release) and, in ventral striatal regions, also via postsynaptic kappa receptors on striatal neurons. These results indicate that dynorphin acts in the striatum to regulate (inhibit) dopamine input to striatonigral neurons. Enkephalin seems to play a similar role for striatopallidal neurons. In these neurons, IEGs are induced by D2 receptor antagonists (neuroleptics). We demonstrated that (1) repeated D2 receptor antagonist treatment produces increased enkephalin expression in striatopallidal neurons which is accompanied by suppressed IEG induction in these neurons; (2) enkephalin (mu and delta opioid receptor) agonists, infused into the striatum, suppress D2 antagonist-induced IEG expression and induce contraversive turning, effects that are dose-dependently blocked by the opioid receptor antagonist naloxone. These results indicate that enkephalin in the striatum inhibits D2 receptor-mediated responses in striatopallidal neurons. Taken together, our results indicate that striatonigral and striatopallidal pathways are under inhibitory control of dynorphin and enkephalin, respectively, and that repeated treatments with psychostimulants or neuroleptics increase this inhibitory regulation of these pathways. Such adaptive responses likely contribute to the behavioral alterations that occur during such drug treatments.

# Symposium 7

## Sym7/1
THE HUMAN VESTIBULAR CORTEX IN PET AND MRI ACTIVATION STUDIES. *M. Dieterich, Th. Brandt, S.F. Bucher. Department of Neurology, Klinikum Grosshadern, Ludwig-Maximilian University of Munich, Marchioninistrasse 15, D-81377 München, Germany*

Several distinct and separate areas of the parietal and temporal cortex have been identified in animal studies as receiving vestibular afferents, such as area 2v at the tip of the intraparietal sulcus, area 3aV in the central sulcus, the parieto-insular vestibular cortex (PIVC) at the posterior end of the insula, area 6 and area 7 in the inferior temporal lobule. Our knowledge about the location and function of vestibular cortex in humans is less precise, derived mainly from stimulation experiments reported anecdotally in the older literature. Thus, functional imaging studies with positron emission tomography (PET) and functional magnetic resonance imaging (fMRI) were recently performed to identify activated cortex areas during different vestibular stimuli.

Caloric irrigation of the ear – physiologically eliciting rotatory nystagmus and vertigo – in PET activates two cortical temporal areas, one in the posterior insula and the other in the anterior insula, as well as an area in the anterior cingulate (Bottini et al., 1994; Dieterich et al., 1996). The activated area in the posterior insula seems to represent the human homologue of the vestibular cortex in monkeys, the PIVC (Grüsser et al., 1990).

Activation of the posterior insula was also unexpectedly found in an fMRI study during horizontal optokinetic nystagmus (OKN) (Bucher et al., 1997). OKN is an integral part of dynamic spatial orientation and elicited by self-motion, object-motion, or both. During OKN activation of multiple, different cortical centers for the control of saccades included all of these areas bilaterally, the parietal eye field (PEF), the frontal eye field (FEF), the prefrontal cortex, and the supplementary eye field (SEF). Furthermore, large parts of the anterior insula were activated during OKN, similar to those activated during caloric irrigation. In contrast, both insular areas – the anterior and the posterior one – were inactive during fixation suppression of OKN. From these activation studies it seems possible that the area in the anterior insula is a relay station for down-stream control of space-oriented eye movements, while the area in the posterior insula is part of the vestibular cortex.

Although all activations during OKN were bilateral, a significant right-hemispherical predominance was found for MT/MST, PEF, FEF, SEF, prefrontal cortex, posterior and anterior insula, and paramedian thalamus (Bucher et al., 1997). Involvement of the vestibular cortex in OKN and the right-hemispherical predominance of the

insula and cortical ocular motor areas was demonstrated for the first time. Functional interpretation refers to the right hemisphere as the dominant for dynamic spatial orientation and the (multisensory) vestibular cortex as the critical site for the evaluation whether OKN is elicited by self-motion or object-motion.

## Sym7/2
CORTICAL CONTROL OF SACCADES. *Bertrand Gaymard. INSERM U 289, Hôpital de la Salpêtrière, 75651 Paris cedex 13 France*

Saccades are controlled by a cortical network composed of several oculomotor areas. It is now well established that the degree of involvement of each of these areas depends on the behavioral context in which the saccade is performed. Our knowledge of the role of each of these areas has greatly improved in the last few years. Functional imaging studies have enabled to localize them with great accuracy, and lesion studies have enabled to better understand their roles.

The parietal eye field (PEF). This area is located in the depth of the intraparietal sulcus. Its efferences travel through the posterior limb of the internal capsule and project on the superior colliculus. This area plays an important role in visuo-spatial integration, and in reflexive visually guided saccade triggering. In humans, a lesion of this area increases saccade latencies and a reduces saccade gain. A lesion involving its efferences selectively impairs reflexive saccade accuracy whereas voluntary saccades are spared.

The frontal eye field (FEF). Recent functional imaging studies have shown that the FEF is localized in the precentral gyrus, immediately rostral to the hand motor area. In humans, a selective lesion of the FEF results in marked saccadic impairments that rapidly recover. Voluntary saccades, especially saccades made in absence of visual target, are markedly impaired. Memory-guided saccades are more impaired when longer delays of memorization are used, suggesting that this area has also memory functions.

The supplementary eye field (SEF). In the monkey, the SEF is located laterally to the pre-supplementary motor area. Functional imaging studies suggest that in humans, it would be located on the medial wall of the frontal lobe, just above the cingulate sulcus. The role of the SEF is not yet fully understood. The only impairments described up to now in patients with an SEF lesion is a difficulty to remember a sequence of saccades and an impairment in the accuracy of memory-guided saccades performed after a body rotation. In monkeys, recent studies suggest that it would play an important role in eye-hand coordination.

The dorsolateral prefrontal cortex (DPFC). This prefrontal area is not involved in saccade triggering per se. However, it has two functions that are closely tight to the oculomotor system. One of these is reflexive saccade inhibition: patients with a selective lesion in the DLPFC are unable to perform an antisaccade task, i.e. a task in which the instruction is to make a saccade away from a suddenly appearing visual target. The other main, well known function of the DLPFC concerns visual short term memory. Patients with a lesion in this area have a selective impairment of memory-guided saccade gain. In primates, this impairment has been shown to increase with longer delays of memorization.

The anterior cingulate gyrus. Recent functional imaging studies have shown that some areas located in the cingulate cortex are activated during various saccade tasks. One of these cingulate areas is located in the posterior part of the anterior cingulate gyrus, a nd has been visualized in several studies. In humans, two patients with an infarct involving this region have been recently studied. The results confirm that the caudal anterior cingulate cortex is involved in saccade control, either through an early activation of the frontal ocular motor areas, or through a direct access to the brainstem premotor structures.

**Sym7/3**

COMBINED DEFICITS OF SACCADES AND VISUO-SPATIAL ORIENTATION AFTER CORTICAL LESIONS. *W. Heide, D. Kömpf. Department of Neurology, Medical University, D-23538 Lübeck, Germany*

Saccadic eye movements are closely linked to visuo-spatial orientation: Functionally, saccades are essential for the exploration of visual space, and a proper cerebral representation of space is a prerequisite for saccadic accuracy. Anatomically, the network of frontal and parietal cortical areas that is known to control saccades also seems to be involved in spatial attention and orientation at least in monkeys. If this is true also for the human cerebral cortex, lesions should cause deficits in both categories. We investigated this in 40 patients with focal unilateral lesions of the posterior parietal cortex (PPC), the frontal eye fields (FEF), the supplementary motor area (SMA) or the dorsolateral prefrontal cortex (PFC). Horizontal saccadic eye movements were recorded by infrared reflection oculography, and visual hemi-neglect or other visuo-spatial disorders were investigated by a series of simple neuropsychological tests such as bisection, cancellation, drawing and copying tasks and the assessment of the internal spatial coordinates (subjective visual vertical and subjective straight ahead). Depending on the site of the lesion, different patterns of deficits were obtained: Lesions of the right more than the left PPC impaired the reflexive exploration of visual space, in terms of delayed and hypometric visually-guided saccades into the contralesional visual hemifield. This was related to the severity of visual hemi-neglect. Further, PPC lesions specifically led to basic deficits of the perceptual analysis of space such as a distorsion of the internal spatial coordinates or an instability of spatial constancy across saccades. The latter was tested by applying double-step stimuli where saccade-related extraretinal information has to be taken into account for achieving spatial accuracy. Saccadic dysmetria in this paradigm correlated with impaired performance in more global tests of visuo-spatial orientation such as copying Rey's figure. In contrast, frontal lesions left these functions intact. FEF lesions, however, impaired the systematic voluntary exploration of space, thus causing an exploratory-motor type of visual hemineglect. Prefrontal (PFC) lesions impaired the working memory for saccade-related spatial information, whereas SMA lesions affected temporal functions such as the timing of saccadic sequences, but did not cause specific visuo-spatial dysfunction. In conclusion, patients with frontal or parietal cortical lesions often exhibit combined saccadic and visuo-spatial disorders. In each case, the identification of the specific pattern of deficits is not only important for the topical diagnosis, but also for an effective rehabilitation.

**Sym7/4**

CORTICAL CONTROL OF SMOOTH-PURSUIT EYE MOVEMENTS. *Thier, U. Ilg, P. Dicke. Sektion für Visuelle Sensomotorik, Neurologische Universitätsklinik, 72076 Tübingen, Germany*

Primates, including humans, make saccades in order to move the images of objects of interest into the fovea and smooth-pursuit in order to stabilize these images on the fovea, in case the object moves at moderate velocity. Both types of goal-directed eye movements have served as useful models of sensorimotor coordination and we have seen considerable progress in our understanding of their neuronal underpinnings, mainly based on experiments on non-human-primates. Saccades as well as smooth-pursuit eye movements depend on visual information processed in the geniculo-cortical pathway and fed into various eye movement related areas located in both the posterior parietal lobe and in frontal cortex. A hallmark of the cortical network, contributing to goal-directed eye movements, is the anatomical segregation into *eye fields* involved in either saccades or in smooth pursuit. The parietal eye fields corresponding to areas LIP, MP, and MST may serve as a case in point. While area LIP, located on the posterior bank of the intraparietal suclus (IPS) and area MP, the latter confined to the medial aspect of the parietal lobe, seem to be largely devoted to saccades, the lateral part of area MST (= MSTl), located on the anterior bank of the superior tempo-ral sulcus (STS), spezializes in pursuit. Also the classical frontal eye field in the arcuate sulcus has recently been shown to consist of anatomically distinct compartments for saccades and pursuit which do not overlap. It is close at hand to speculate that the existence of anatomically segregated sets of cortical eye fields for saccades and pursuit reflects different phylogenetic backgrounds of these two types of goal-directed eye movements. Why are there multiple cortical eye fields? We argue that the multiplicity of eye fields indicates functional specialization rather than redundancy, an idea which can be easily exemplified by looking at the parietooccipital areas involved in pursuit. Area MT/V5, located in the posterior bank of the STS is an extrastriate area involved in the analysis of visual motion for many purposes, both perceptual and visuomotor, including the extraction of target motion for pursuit. MT/V5 is as yet the only pursuit-related cortical structure whose human homologue has been defined with sufficient precision based on myeloarchitectonic criteria and functional imaging. Unlike MT/V5, the lateral part of neigboring area MST (MSTl) should be looked upon as a supramodal cortical area, integrating signals on retinal image slip with (non-visual) information on eye and head movement and top-down influences needed in order to define a target and to decide what target among many to pursue. The idea that MSTl offers a distributed representation of target-motion in space, useful for the guidance of eye movements as well as other types of goal-directed movements is based on this view and has proven to be profitable for our interpretation of pursuit deficits following parietal lesions in animals and humans. Finally, pursuit-related signals found in the adjoining dorsal part of MST (MSTd) most probably do not indicate a contribution to the control of smooth-eye movements but an important role of pursuit-related signals in the processing of the visual consequences of ego motion needed for the detection of heading as well as the maintanance of perceptual stability despite self-induced retinal image slip.
Supported by the Deutsche Forschungsgemeinschaft and the EU HCM programme.

## Symposium 8

**Sym8/1**

ANISOMETRY OF SPACE REPRESENTATION IN PATIENTS WITH UNILATERAL NEGLECT. *E. Bisiach, M.R. Colombo, G. Geminiani. Dipartimento di Psicologia, Università di Torino, Via Lagrange 3, 10123 Torino, Italy*

Patients suffering from left neglect typically bisect horizontal lines to the right of the true midpoint under visual control. When asked to mark the endpoints of a virtual line of a given length on the basis of its midpoint printed on a sheet of paper, the distance they generate between the left endpoint and the midpoint is usually larger than the distance they generate between the latter and the right midpoint. Similarly, when they are asked to extend leftwards a horizontal segment of which the left endpoint is located in the middle of a sheet of paper so as to double its length, the segment they generate is usually longer than the segment they generate in the converse (rightward) condition. In both cases, the relative error (i.e. the ratio between the left and right virtual or actual segments drawn from the landmark given as midpoint) has the same direction (i.e. rightward) typically found in these patients with canonical line bisection tasks. It has been argued that this finding cannot be accommodated by current interpretations of unilateral neglect. It has also been suggested that is is indicative of a left-right anisotropy of the medium underlying visual space representation. The problem still remains as to the extent to which the observed behaviour is specifically dependent on the involvement of the visual modality. The results of an experiment aimed at assessing the relative role of visual and haptic modalities suggest that the behaviour in question may be found in both modalities, though perhaps to a different degree and subject to double dissociation.

**Sym8/2**

COMPONENTS OF UNILATERAL NEGLECT AFTER RIGHT-HEMISPHERE LESIONS. *J. Driver. Department of Psychology, Birkbeck College, University of London, Malet Street, London WC1E 7HX, UK*

There have been many recent suggestions that unilateral neglect is a sydrome which can involve several different component deficits. A series of experiments seeks to identify and separate some of these components. Previous attempts to separate sensory versus motoric components of neglect may have been confounded by general difficulties with incompatible tasks after brain damage. A reaction time methodology is proposed for separating lateral biases in vision from biases in the preparation of reaching movements with the ipsilesional arm, without introducing incompatibility into the task. The results imply that patients with right inferior-parietal lesions have both visual and motoric biases, whereas the motoric bias is absent in neglect patients with more anterior lesions, contrary to recent claims for the opposite association with lesion site. Further experiments examine lateral biases in visual perception, versus in short term visual memory, and find that the former are associated more with parietal lesions, and the latter with more anterior lesions. A final series of experiments examines the possible role of 'body image', and of the associated updating of egocentric position, in tactile aspects of neglect. Tactile extinction in a group of right-hemisphere patients, under concurrent stimulation of the two hands, is found to be more severe when the unseen hands are placed further apart, confirming that the defict arises at a spatial level of representation where proprioception signals the current body posture.

**Sym8/3**

ENCODING OF SPACE IN PATIENTS WITH NEGLECT. *Hans-Otto Karnath. Department of Neurology, University of Tübingen, Hoppe-Seyler-Str.3, D-72076 Tübingen, Germany*

The characteristic disturbance of patients with neglect is a failure to orient towards and to explore the contralesional part of space by eye or limb movements. It has been argued that an altered representation of egocentric space might underlie this disturbance. The actual *gestalt* of this representation, however, remained unclear. Different hypotheses were put forward. Suggested were a *rotation* of the whole egocentric reference frame around the earth vertical body axis or a *translation* of the reference system toward the side of the lesion. Further proposals were a *uniform compression* of space representation, a *shrinkage* only on the contralesional side, or an extension on the contralesional and compression on the ipsilesional side according to a *logarithmic transformation*. We tested these different hypotheses in patients with left-sided neglect who had no visual field defects. To distinguish between the rotation and translation hypotheses, perception of "straight ahead" body orientation was measured by presenting a red LED at two different distances away from the subject's body. At both locations, the subjects' task was to direct the LED to the position which they felt lay exactly "straight ahead" of their bodies' midsagittal plane. We found that the egocentric deviation was clearly due to a *rotation* of the reference frame around the earth vertical body axis to the ipsilesional side. In a second experiment, the subjects' perception of subjective equidistance was determined in the horizontal plane. To distinguish between the different hypotheses suggesting an anisometry of space representation in the horizontal plane, subjects had to position ten red LEDs equidistantly along a semicircle, which was positioned horizontally in front of them at eye level. Half of these LEDs had to be positioned on the subjects' left, the other half on their right side. No anisometry was observed that could account for the patients' failure to orient towards and to explore the contralesional part of space. In conclusion, the findings favour a disturbed input transformation in patients with neglect that leads to a deviation of the whole egocentric space representation toward the ipsilesional side.

**Sym8/4**

VISUAL SIZE PROCESSING IN SPATIAL NEGLECT. *AD Milner & M Harvey*. School of Psychology, University of St Andrews, St Andrews, Fife KY16 9JU, UK. * Now at Dept of Psychology, University of Bristol, Bristol BS8 1TN, UK*

Evidence will be presented from the use of the Landmark task and a size matching task that patients with left-sided neglect systematically under-perceive visual extent in leftward parts of space. It is suggested that this helps to explain the occurrence of rightward line-bisection errors in most neglect patients. Evidence will also be presented that apparently comparable errors in size perception can be induced as a function of spatial cueing in normal subjects. It is tempting to infer from these two sets of findings that the second may explain the first, that is that attentional biases of a chronic nature may underlie the perceptual effects seen in our neglect patients. The pattern of results when one or other end of the lines used in the Landmark task is cued, however, is exactly opposite to what such a hypothesis would predict. It will be argued, therefore, that a different kind of explanation of the size processing distortions seen in neglect is needed.

## Symposium 9

**Sym9/1**

THE FUNCTIONAL ANATOMY OF SYNTACTIC PROCESSING: EVIDENCE FROM BRAIN-IMAGING RESEARCH. *C.M. Brown[1], P. Indefrey[1], P. Hagoort[1], H. Herzog[2], R. Seitz[3]. [1] Max Planck Institute for Psycholinguistics, "Neurocognition of Language Processing" Research Group, Wundtlaan 1, NL-6525 XD Nijmegen, The Netherlands. [2] Institut für Medizin, Forschungszentrum Jülich GmbH, D-52425 Jülich, Germany. [3] Neurologische Klinik der Medizinischen Einrichtungen der Heinrich-Heine-Universität Düsseldorf, Postfach 10 10 07, D-40001, Düsseldorf, Germany*

Syntactic processing is a core feature of human language competence. The ability to assign grammatical roles to individual words and groups of words is a critical part of achieving coherence in both language production and comprehension. The central role of syntactic competence is highlighted in brain-damaged populations, where grammatical disturbances can lead to a partial or even total loss in processing language. While this makes it important to identify the neural structures that support syntactic processing, such structures are at the same time difficult to identify because syntactic processing never operates in isolation: Normal language processing requires the orchestration of many different sources of linguistic and non-linguistic knowledge. We will report on PET brain-imaging work in which we minimized the influence of non-syntactic information, by presenting subjects with a task that required the production of grammatical sentences consisting of pseudowords and real function words. One of our main findings indicates a specific role during syntactic processing for the dorsal border of Broca's area. This area can be distinguished from areas activated by other language processes, as well as from areas activated by explicit judgement tasks, which have been used in other brain-imaging studies of sentence processing.

**Sym9/2**

A MULTI-MODALITY APPROACH TO THE FUNCTIONAL NEUROANATOMY OF AUDITORY. LANGUAGE PROCESSING. *J-F Demonet. INSERM U 455, Hopital Purpan, 31059 Toulouse cedex, France*

In previous Positron Emission Tomography (PET) studies in normal right-handed volunteers, we compared brain activations elicited by either phonological or lexical semantic monitoring tasks and described differential distributions of increases in regional cerebral blood flow over cerebral hemispheres as phonological processes induced activations near the left sylvian fissure while lexical semantic tasks elicited a widely distributed pattern involving association cortical areas of temporal, parietal, and frontal lobes in both hemispheres. As PET provides only averaged data on brain regions activated over one minute, the same paradigm was used to further explore these language-specific neural correlates in terms of space and time resolution. Functional MRI using multislice echoplanar sequences was used to replicate this paradigm and assess the influence of subjects' handedness and gender on brain activations. Multi-channel Evoked Related Potentials using Neuroscan was used to explore in the time domain the neural counterparts of these monitoring tasks that both involved sequential processing of cues and targets versus distractors. The phonological task consisted of detecting phoneme /b/ in the last 2 syllables of 4 (CV)-syllable pseudowords if and only if phoneme /d/ was present in the first 2 syllables; 4 types of stimuli were therefore generated, consisting of 2 varieties: 2 cued types (/d/ – /b/ (target), /d/ – x) and 2 uncued types (x – /b/, x – x). The same applies to the lexical semantic task using adjective-noun pairs in which targets were names of small animals (e.g. mouse) if and only if preceded by positive adjective (e.g. kind), other stimuli types being a cued type (positive adjective paired with big animal), and 2 uncued types (negative – small, and negative – big). After an initial common phase (N1, P2), ERPs showed that the 2 types of stimuli generated waves that splitted apart in the same way in both tasks with cued stimuli (possibly involving targets) generating negative shifts while uncued stimuli elicited positive shift. However this splitting point, probably reflecting either phonological or semantic decision, (a) occurred after the end of the 2nd syllable in pseudowords, versus before the end of the acoustic trace of adjective, suggesting a step-by-step sequential processing in the phonological task in contrast with accelerated, probabilistic lexical processing in the semantic task and (b) corresponded to distinct spatial distributions of electrical potentials over the scalp reflecting generators located by Brain Electrical Source Analysis in the left hemisphere for the phonological task and in both hemispheres for the lexical semantic task, in good accord with previous results on PET data. In general the combination of imaging techniques providing spatial resolution on the one hand and temporal resolution on the other seems a powerful way to improve our knowledge of the spatiotemporal dynamics of large-scale neural ensembles that subserve language functions; our data evidenced (at least) two interconnected subsystems, one being associated with phonological processing spatially distributed around the left sylvian fissure and operated in a sequential mode, the other, related to lexical semantic processing, being more widely distributed throughout the entire brain and operated in a paralell mode.

## Sym9/3

BRAIN POTENTIALS ELICITED BY SYNTACTIC REVISION PROCESSES: *A. Mecklinger & A.D. Friederici: Max-Planck-Institute of Cognitive Neuroscience, Inselstrasse 22-26, 04103 Leipzig, Germany*

A large variety of studies using event-related potential (ERPs) as on-line measures of language processing have focused on processing lexical/semantic information. In recent years, however, a couple of studies also examined ERP indices of syntactic processes. These studies consistently report a late positivity which is either evoked by outright syntactic violations or by violations of structural preferences requiring syntactic revision or reanalysis processes. A series of experiments will be reported in which the functional significance of the late positivity during syntactic revision processes was examined. Participants read sentences with either relative clause or complement clause structures which were subject-first/object-first ambiguous unless they are disambiguated in mid-sentence or sentence final positions. Based on recent parsing models the revision processes can be considered to be less complex for relative clause structures than for complement clause structures. The results consistently show that encountering the less-preferred object-first structure in either mid-sentence or sentence-final positions evokes a late positivity. In showing an earlier onset latency for relative clause structures than for complement clause structures this component's latency apparently is associated with the complexity of the required reanaly-

zes. The results indicate that (a) initially ambiguous structures are revised as soon as disambiguation information is available and (b) that the latency of the late positivity reflects some aspects of the complexity of these on-line revision processes.

## Sym9/4

DEFICITS OF SEMANTIC MEMORY AND NAMING IN PROGRESSIVE FLUENT APHASIA. *K. Patterson[1] and J. R. Hodges[1,2]. [1] Medical Research Council Applied Psychology Unit 15 Chaucer Road, Cambridge CB2 2EF, UK. [2] Department of Neurology University of Cambridge Clinical School Addenbrooke's Hospital Hills Road, Cambridge CB2 2QQ, UK*

Progressive fluent aphasia, resulting from focal temporal lobe atrophy, affects lexical and semantic components of language; phonological and syntactic aspects of speech production are relatively preserved, at least until very late in the course of the disease. The most striking deficit in speech production is a profound disruption of the ability to produce specific content words. This anomia typically affects production of proper names (people and places), common names (objects), and also specific verbs which are often replaced in spontaneous speech with more general verbs (e.g., "make" or "do" in English). The goals of the research reported here include: (i) specification of the brain regions crucial for comprehension and production of content words, as revealed by structural and functional brain imaging with patients and also functional activation studies with normal subjects; (ii) analysis of detailed longitudinal patterns of semantic deterioration as evidence for the organisation of conceptual knowledge; and (iii) analysis of detailed longitudinal patterns of anomia as evidence for models of speech production.

## Symposium 10

### Sym10/1

AN EVOLUTIONARY APPROACH TO SPATIAL MEMORY AND THE HIPPOCAMPAL CHALLENGE: FROM MICE AND MONKEYS, TO MARSH TITS AND MAN. *Nicola S. Clayton. Section of Neurobiology, Physiology & Behavior, Briggs Hall, University of California Davis, Davis CA 95616, USA*

Comparative studies provide a unique source of evidence for the role of the hippocampus in learning and remembering information about spatial locations. The hippocampal volume of a scatter-hoarding species that caches food in many different locations is larger, relative to the remainder of the telencephalon, than that of a species which caches all its food in one larder, or does not cache at all. Several scatter-hoarding species have a more accurate and enduring spatial memory, or a preference to rely more heavily upon spatial cues, than that of closely related species which store less food, or none at all. But why do scatter-hoarding species have an enlarged hippocampus and a better spatial memory or a preference to rely more upon spatial than non-spatial cues? What is the mechanism by which the hippocampus learns and remembers spatial locations, and might the mechanism differ between birds and mammals ? The current mammalian dogma is that the neural mechanisms of learning and memory are achieved by variations in the number and effectiveness of connections or synapses, which are associated with a relatively fixed corpus of processing units, the neurons. By contrast, the avian brain has evolved a novel mechanism of varying the number of processing units in response to mnemonic demand. Recent work on the dual ontogeny of food-storing and the avian hippocampus illustrates this dynamic interaction between brain and behavior: some aspect of the experience of remembering the spatial location of food caches triggers dramatic changes in size and neuron number, as well as in rates of cell birth and death.

**Sym10/2**

THE DELAY-BRION SYSTEM AND THE EXPLANATION OF AMNESIA. *David Gaffan. Department of Experimental Psychology, Oxford University*

In monkeys, any interruption of the Delay-Brion system, a connected system of structures including the hippocampus, fornix, mamillary bodies and anterior thalamus, produces an impairment in memory for unique complex scenes. This scene memory impairment is functionally similar to human episodic memory impairment, since normal episodic memory frequently involves the reconstruction of a remembered scene. Patients with discrete fornix transection following colloid cyst removal have clinically significant impairments in episodic memory, which can be measured with delayed paragraph recall or delayed reproduction of the Rey figure. However, these patients are not densely amnesic. They can score in the normal range in the Warrington Recognition Memory Test, for example. Thus, although the scene-memory function of the Delay-Brion system is an important component in human memory, it is not the whole explanation of amnesia. Dense amnesia is caused in monkeys by disconnection of subcortical intrahemispheric interactions between the temporal lobe and the frontal lobe. For example, crossed unilateral ablations of the perirhinal cortex in one hemisphere and the frontal lobe in the other produced a severe impairment in visual recognition memory tested by delayed matching-to-sample with objects in a constant background scene, a task similar to the Warrington Recognition Memory Test for faces. Some possible subcortical routes of interaction between the temporal and frontal lobe outside the Delay-Brion system, and their possible roles in the explanation of human amnesia, will be discussed.

**Sym10/3**

ON THE SPATIAL INFORMATION USED BY THE NEURAL INFORMATION PROCESSING SYSTEMS. *B. Poucet. Center for Research in Cognitive Neuroscience, CNRS, Marseille, France*

Behavioural work suggests that the reference frame used by rats to orient in space and to memorize locations is based on a selective spatial information-processing system. Distant landmarks and global configurations are preferred over local cues and featural details. Recordings of the neuronal unit activity in the dorsal hippocampus of freely moving rats suggest that the hippocampal place cell system can be also described as a selective spatial information-processing system. Overall surfaces and configurations of relatively distant landmarks exert more powerful control over place cell activity than featural cues in proximal space. In addition both spatial behaviour and hippocampal place cell activity are strongly dependent upon an animal's ability of monitoring its orientation. This sense of orientation is provided by a system that processes directional information so that the animal knows which direction it faces at each time. One locus where directional information might be processed in the brain is the postsubiculum which contains head direction cells. Positional (place-cell system) and directional (head-direction system) computations depend on the use of both static (essentially visual) and dynamic, movement-related cues (i.e., vestibular and kinaesthetic cues). If we assume that both positional and directional computations occur within the reference frame provided by the visual environment, then the spatial information used by the two systems is necessarily constrained by the same structural properties of space, and must satisfy the constraints of both types of computation. More specifically, positional computations are made easier when movements result in large variations in apparent visual angle between landmarks. Thus relatively close landmarks are more informative than more distant landmarks. In contrast, directional computations are made more reliable when parallax effects are minimized, thus when reference landmarks are distant. These contradictory requirements might have constrained the neural systems for processing spatial information to evolve so that they preferentially select mid-distance landmarks that satisfy both positional and directional constraints.

**Sym10/4**

A PLACE FOR OLFACTORY CUES IN SPATIAL ORIENTATION BY YOUNG AND OLD RATS: *F. Schenk, C. Barras & P. Lavenex: Institut de Physiologie, Université de Lausanne, Rue du Bugnon 7, CH-1005 Lausanne, Switzerland*

Although nocturnal, adult rats ignore olfactory cues for orientation and rely preferentially on distant visual cues. In a first series of experiments we have shown that adult rats disregarded olfactory cues for accurate choice in an illuminated radial maze whether or not distant visuospatial cues were available. In darkness, the same olfactory cues were used predominantly, as if released from a light induced inhibition.

Anatomically, olfactory and visuospatial sensory information reach the hippocampus through differently organized access pathways. Vision has become dominant for orientation through the evolution of vertebrates as it provides more accurate and stable distant information. However, olfaction plays an important role in immature animals. A second series of experiments has analyzed when and how visuospatial cues become dominant over olfactory cues through the life span. A homing task was used to test spatial memory in immature subjects as well. Following training to reach home via an escape hole at a fixed position in space, rats had to choose between the spatial position defined by distant visual landmarks and another position marked by local olfactory traces. Very immature rats (18 days PN) were concentrated on the olfactory defined position. From 21 days, a primitive spatial representation was based on the integration of movement derived cues with olfactory marks. Though unable to perform place learning, immature rats showed a conflict between the spatially defined and the olfactory escape positions following training with associated spatial and olfactory cues. This indicates that polymodal sensory information including olfactory cues might play a critical role in the development of an early spatial representation. Aging processes appear to affect this mechanism of movement memory hypothetically based on path integration, with the consequence that 12 month old rats appear more dependant than younger adults on a reference memory of local cues to discriminate the training position. In the light, this memory would be akin to a visual snapshot of the panorama surrounding the goal area, hence the preference for the spatial position in conflict trials. In darkness this reference memory would be olfactive, hence the preference for the "olfactory place" in conflict trials.

This analysis of the primary orientation abilities shown by immature rats provides a model of how sensory information is integrated into a spatial representation. It suggests also that aging effects on spatial memory are already evident in the second year.

## Symposium 11

**Sym11/1**

HUMAN SOMATOMOTOR CORTICAL RHYTHMS AND THE TIMING OF MOTOR CONTROL. *R. Hari. Brain Research Unit, Low Temperature Laboratory, Helsinki University of Technology, Box 2200, FIN-02015 HUT, Finland*

The human somatomotor cortex displays prominent rhythmic oscillations around 10 and 20 Hz. The 10-Hz rhythm seems to receive its major contribution from the postcentral somatosensory cortex whereas the 20-Hz rhythm clusters slightly more anterior, suggesting generation in the precentral motor cortex. The 20-Hz rhythm is strongly enhanced after voluntary movements, with the generation sites following in a somatotopical manner the body part moved. Similar 20-Hz enhancements are also observed after electric stimulation of the peripheral nerves.

The periodicity of precentral rhythms may in some conditions be closely related to timing of motor unit firing in the moving muscle. For example, the 15–33 Hz MEG signals originating in the precentral motor cortex were found to be coherent with motor unit firing of isometrically contracted muscles. The sites of origin of the coherent signals were close to the hand motor area for all upper limb muscles studied, with no systematic intermuscle differences, and close to the foot motor cortex for lower limb muscles. The cortical signals preceded the motor unit firing by 12–53 ms, depending on the musle. We suggest that the cortical rhythms reflect a common corticospinal drive to the motoneuron pool. This drive, although evidently forming only a tiny portion of all neuronal commands reaching the motor units, may still be significant in shaping the timing of motor unit firing and in optimizing motor cortex ouput.

Hari R, Salmelin R. Human cortical rhythms: a neuromagnetic view through the skull. *Trends Neurosci* 1997, 20: 44–49.

Salenius S, Portin K, Kajola M, Salmelin R, Hari R. Cortical control of human motoneuron firing during isometric contraction. *J Neurosci* 1997, in press.

Salmelin R, Hämäläinen M, Kajola M, Hari R. Functional segregation of movement-related rhythmic activity in the human brain. *NeuroImage* 1995, 2: 237–243.

## Sym11/2

TEMPORAL REPRESENTATIONAL CHANGES WITH LEARNING. *Michael M. Merzenich. Sooy Professor, Keck Center, University of California at San Francisco, San Francisco CA 94143-0732*

The cerebral cortex is a "learning machine" that remodels itself in detail throughout life. Hebbian and competitive Hebbian network effects account for basic input selection, input integration and input segregation phenomena that are recorded in the cortex in learning. The fact of plasticity infers that representation in the cerebral cortex must be relational. The creation of cell assemblies in Hebbian plasticity is the presumptive basis of generation of relational representations; their creating and strengthening recorded in progressive learning experiments result in the emergence of more strongly temporally correlated responses and consequently in increased representational salience from behaviorally engaged cortical networks.

Changes induced by learning are strongly modulated by ACh inputs from the basal nucleus of Meynert. ACh is normally released only during attended behaviors. Appropriately temporally modulated inputs from that nucleus alone are sufficient to enable at least most of the enduring changes recorded in learning-induced cortical plasticity experiments. Other regulatory transmitters are, of course, also involved in the modulation of enduring brain changes in learning and memory. In perceptual, cognitive or motor skill learning across childhood, the brain constructs an increasingly wider and deeper platform of well-learned and automatically performable behaviors. On this growing platform of automatic behaviors, the brain is continually differentiating its representations as new behavioral demands arise. Top-down inputs signalling on-going "expectation" or "memory" are a critical participant in these processes.

These dynamic processes have now been demonstrated to contribute to the origins of and the phenomenology of neurological disabilities, e.g., to the origins of occupationally based movement disorders and of language-based learning disabilities. Understanding of these brain processes have led to successful methods for ameliorating these and potentially other widely occuring neurological problems. Some of the main findings from these pragmatic studies of brain plasticity and learning will be briefly summarized.

Supported by NIH Grant NS-10414, the Coleman Fund, and HRI.

## Sym11/3

THE ROLE OF TIME STRUCTURE IN SENSORY RESPONSES FOR ASSOCIATIONAL PLASTICITY IN LARGE POPULATIONS OF NEURONS IN THE NEOCORTEX. *Matthias H.J. Munk, Max-Planck-Institut für Hirnforschung, Deutschordenstraße 46, D-60528 Frankfurt, Germany*

Information processing in the neocortex has to cope with combinatorial complexity in the feature space of sensory and motor actvity. As the cortex has to perform as rapidly and as reliable as possible, coding of information must be based on a flexible and redundant mechanism which in addition requires the ability to store the relations among components of neurally coded information. Explicit coding in the form of responses of individual, sharply tuned neurons would require more cells than are available and would not allow for rapid regrouping. If instead coding is performed by groups of neurons cooperating as assemblies, using individual cells in different combinations for the representation of different contents, the number of required representational elements is much reduced. Such relational representations are highly dynamic and therefore allow for the required coding flexibility. Our current hypothesis is that assemblies are defined by the synchronicity of their firing. If these assemblies transmit synchronized signals to neuron populations at subsequent processing stages, the strength of synchronization will determine how efficient postsynaptic responses will summate and how much of these signals will then influence postsynaptic spike responses. According to this hypothesis, two assemblies signalling the presence of two separate objects, should not fire in synchrony. However, learning processes should be able to modify the representation of two objects if they become perceived as belonging together. Our prediction is that the two previously non-synchronized assemblies get fused. We directly tested the idea whether functional assemblies of cortical neurons can be temporarily linked by conditioning sensory responses in visual cortex with experimental activation of neuromodulatory systems: electrical stimulation of the Mesencephalic Reticular Formation (MRF) immediately before presenting a visual stimulus reliably induced oscillations in multi-unit spike responses, typically at frequencies around 30@emsp14;Hz. We succeeded to fuse two assemblies temporarily, i.e. the oscillatory responses synchronized across different groups of cortical neurons when they were visually coactivated and conditioned with MRF stimulation. Most importantly, when conditioned for about 300 trials, such newly formed assemblies continued to fire synchronously for up to 45 minutes after the MRF activation was discontinued. This associational representation could be experimentally interchanged among different groups of neurons. We conclude that functional assemblies defined by the synchronicity of their spike firing can be reversibly linked if during simultaneous sensory stimulation, neuromodulatory systems like the MRF become active. In addition it seems likely that MRF-induced gamma oscillations play an important role for fusing assemblies and may even help to enable synaptic plasticity.

## Sym11/4

TEMPORAL CONSTRAINTS OF COGNITION: FURTHER EVIDENCE FOR TEMPORAL PROCESSING ON THREE DIFFERENT TEMPORAL LEVELS: *N. v. Steinbüchel: Institute f. Med. Psychology , Goethestr. 31, D-80336 Munich*

On the basis of neuropsychological, psychophysical and brain imaging findings it can be concluded that sensory information is processed in a discrete fashion. In this paper experimental evidence on temporal constraints of information processing on three different temporal ranges in brain injured patients with focal lesions in the left and right hemisphere, reading and writing impaired children and healthy control subjects is reported. On a high-frequency level, atemporal system states with the duration of approx. 30 ms – implemented by neuronal oscillations – are conceived of as providing the logistical basis for the identification of elementary events and succession. Language skills seem to be strongly associated with this temporal mechanism, which can be assessed with the order threshold paradigm. Here the minimal time interval necessary to indicate the correct temporal order of two acoustic stimuli is measured. Only the order threshold measurements of patients with Aphasia (predominantly Wernicke Aphasia) and children with reading and writing impairments show prolongations up to and over 200 ms, in comparison to older healthy control values around 50–60 ms and control children values of approx. 100 ms.

On another hierarchically distinct level, successive system states are automatically linked together. This temporal integration appears to be limited to intervals up to approx.

2,5–3 secs. Experimental evidence for this presemantic integration comes from a number of different paradigms. For instance in sensorimotor synchronization subjects can anticipate stimulus occurence up to approx. 2,5–3 secs, but not beyond. This temporal range is assessed with different paradigms like temporal reproduction of optic and acoustic stimuli and spontaneous alteration rates of visual and auditory ambiguous figures. Here only patients with frontal lobe lesions show altered temporal behaviour.

Between these two temporal levels in the range of 150–500 ms temporal aspects of movement control are located. This level is assessed with the tapping paradigm, were again only patients with Aphasia (predominantly Broca Aphasia) show selective alterations in central timing.

## Symposium 12

### Sym12/1
THE ROLE OF IONOTROPIC GLUTAMATE RECEPTORS IN LEARNING AND MEMORY – STATUS QUO OF PHARMACOLOGICAL APPROACH. *W. Danysz, W. Zajaczkowski, T. Frankiewicz, and C.G. Parsons. Dept. Pharmacol. Merz+Co, Eckenheimer Landstrasse 100, 60318 Frankfurt /M, Germany*

Over the last 11 years considerable evidence has accumulated indicating the involvement of NMDA receptors in learning processes. This involves both electrophysiological and behavioural studies using mainly pharmacological approaches. In turn, it is generally accepted that the blockade of NMDA receptors leads to learning impairment while their stimulation, or positive modulation may produce enhancement of learning. However, depending on the model used, exactly the opposite effect can also be seen. We observed that in rats after entorhinal cortex lesions, the uncompetitive NMDA receptor antagonist memantine decreased the frequency of reference memory errors in the radial maze paradigm, but the glycine site partial agonist d-cycloserine was without effect. In contrast, in the passive avoidance test, d-cycloserine, but not memantine attenuated anterograde amnesia produced by scopolamine. Also in the same paradigm, NMDA itself (25–100 mg/kg) induced amnesia that was attenuated by MK-801 and memantine. Similarly, in hippocampal slices 10 µM NMDA depressed AMPA receptor-mediated fEPSPs in the CA1 region and also caused a moderate reduction of LTP induction/expression and this later effect was attenuated by the NMDA receptor antagonist memantine. Moreover, although systemically-active antagonists selective for the glycine site of the NMDA receptors produced an apparent impairment in a passive avoidance test they had no negative effects on radial maze learning. Finally, in view of the essential role of AMPA receptors in glutamatergic transmission, the fact that the AMPA receptor antagonists, NBQX and GYKI-52466 have no effect on learning in a number of paradigms (e.g. passive avoidance, T-maze, radial maze) is very surprising. This diversity of effects indicates that a delicate, physiological balance of AMPA and NMDA receptor activation is a prerequisite for optimal functioning of learning processes and that the effect of NMDA receptor antagonists depends on the test used.

### Sym12/2
GLUTAMATE AND DOPAMINE RECEPTORS INVOLVEMENT IN THE MODULATION OF MEMORY PROCESSES. *A.Mele[1], P.Roullet[2], M.Ammassari-Teule[3] and A.Oliverio[1,3]. [1] Dip. Genetica e Biologia Molecolare, Università di Roma "La Sapienza", Ple. Aldo Moro 5, 00185 Roma. [2] ATIPE Cognisciences, Institut des Neurosciences, CNRS, Paris France. [3] Istituto di Psicobiologia e Psicofarmacologia, C.N.R., via Reno 1, 00198 Roma*

It has been suggested that glutamatergic and dopaminergic neural systems interact within the nucleus accumbens (N. Acc.) to interface biologically relevant informations with motor output. Purpose of this study was to investigate the role of these two neurotransmitter systems in the capability of mice to encode spatial and non spatial information. The task consists in placing mice in an open field containing five objects and, after three sessions of habituation, examining their reactivity to object displacement (spatial change) and object substitution (non-spatial change). Any renewal of exploration towards a displaced or a new object is assumed to rely on a comparison between the current situation and a "stored representation" of the inital one.

In a first series of experiments, the effect of systemic administration of the non-competitive NMDA antagonist (MK-801, 0.1; 0.25 mg/kg), as well as the D1 (SCH23390, 0.01; 0.015 mg/kg) and the D2 (haloperidol, 0.04; 0.08 mg/kg) dopaminergic antagonists was studied in CD1 mice placed in the above described situation. The results show that the NMDA antagonist, at low doses, impaired selectively the reactivity of mice to spatial change without affecting any other behaviour but, at higher doses, decreased reactivity to both spatial and non-spatial change as did dopaminergic antagonists.

The second series of experiments was aimed at focusing more specifically on the role of N. Acc. NMDA receptors. Local injections of both the competitive, AP5 (0.1 and 0.15 mg/side), and the non-competitive NMDA antagonists, MK-801 (0.15 and 0.3 mg/side) were performed in mice tested in the same behavioral paradigm. In general, the behavioural effect of intra-accumbens injections of AP5 and MK-801 resembled closely those of systemic administrations of MK-801. That is, NMDA antagonists, at low doses, produced a selective impairment of the reactivity to spatial change but, at higher doses, were found to affect a large range of behaviours among which locomotor activity, habituation and reactivity to spatial and non spatial change.

Taken together these results indicate that NMDA receptors can play a rather selective role in spatial information encoding. In particular, the fact that both systemic and intra-accumbens injections of NMDA antagonists were found to produce a dose-dependent selective impairment of reactivity to spatial change indicates that NMDA receptors located in the N. Acc.mediate, at least partly, operations of spatial information encoding. A possible mechanism for this impaired reactivity to spatial change may be a block of the information flow from limbic and cortical afferents to the N. Acc. Conversely, dopamine antagonists as high doses of NMDA antagonists do not appear to exert any selective behavioural effect. Rather, the locomotor activity effects and object exploration scores recorded in mice with such treatments indicate consistent motor alterations with possible motivational and attentional deficits.

### Sym12/3
CRITICAL INVOLVEMENT OF METABOTROPIC GLUTAMATE RECEPTORS IN LONG-TERM POTENTIATION AND MEMORY FORMATION. *K. G. Reymann. Federal Institute for Neurobiology and Research Institute of Applied Neurosciences, Brenneckestr. 6, D-39118, Magdeburg, Germany, E-mail: reymann@ifn-magdeburg.de*

Both long-term potentiation (LTP) and memory formation are multistage processes. According to our three-stage model of LTP, the $Ca^{2+}$/Calmodulin-dependent short-term potentiation (STP) is followed by a protein kinase C (PKC)-dependent LTP1 stage and a late protein synthesis-dependent LTP2 stage. NMDA receptor activation during tetanization is an essential condition for all 3 stages of potentiation following 100 z tetanization. Additionally a co-activation of metabotropic glutamate receptors (mGluR) is necessary for mechanisms enabling the late stages/maintenance of LTP.

We investigated field potentials-LTP of the CA1 and dentate gyrus of rats both in vitro and in vivo. mGluR antagonists block LTP1 and 2, not interfering substantially with STP. Interestingly, the same antagonists prevent memory formation in a hippocampus-dependent spatial learning task (modified Y-maze). Using a higher tetanus frequency (400 Hz) in the dentate gyrus we observed a form of LTP which is mGluR-, but not NMDAR-dependent. Investigation of new mGluR agonists/antagonists selective for class I led us to the conclusion that phospholipase C-coupled class I mGluRs are critically involved in STP-LTP transformation and consolidation of long-term memory. Class II mGluRs rather inhibit LTP, but might be crucial for long-term depression. Class III mGluRs appear to play a role in regulation of basal synaptic transmission, and thereby can regulate the thresholds for LTP and LTD.

Recent evidence obtained with recordings of single CA1 pyramidal cells indicates that STP and LTP are maintained at least partially by a postsynaptic mechanism since the sensitivity of potentiated neurons to test pulses of the iontophoretically-applied glutamate receptor agonist AMPA increases in 1–2 min after tetanization in both NMDA-, mGluR- and kinase-dependent manner.

Taken together, depending on tetanus parameters both NMDARs- and class I mGluRs are important for LTP-induction, whereas the expression of LTP is realised via AMPARs. Similar mechanisms might be involved in memory formation of hippocampus-dependent spatial learning.

## Sym12/4

ARE THE GLUTAMATE RECEPTORS SPECIFICALLY IMPLICATED IN SOME FORMS OF MEMORY PROCESSES? *A. Ungerer, C. Mathis, and C. Melan. Laboratoire de Psychophysiologie, ULP, URA 1295 CNRS, 7 rue de l'University, 67000-Strasbourg, France*

Considerable evidence indicates that glutamatergic neurotransmission is involved in biochemical events underlying learning and memory processing. In support of this, N-methyl-D-aspartate (NMDA) receptor antagonists disturb acquisition and retention in various learning tasks, suggesting mediation by NMDA receptors of learning and memory processes, which in turn appears to be closely linked to the role of these receptors in synaptic plasticity of the central nervous system. On the other hand, recent studies indicate that metabotropic glutamate receptors (mGluRs) coupled to G-protein are critically involved in synaptic plasticity of various brain structures, thus suggesting that may have an essential role in some learning and memory processes. However, the effects of the substances interacting with NMDA or mGluRs on learning and memory processes widely differ in their amplitude and their time-course according to the learning task used. Thus, NMDA receptor antagonists do not affect acquisition processes per se, or retrieval, but appear to interfere specifically with the formation of memory traces. However, this action is highly task-specific. Indeed, we observed that systemic injection of the competitive NMDA receptor antagonists, gamma-L-glutamyl-L-aspartate (gamma-LGLA) and CPP, or ICV injection of D-AP5, immediately following acquisition of a Y-maze avoidance learning task in mice, deeply impaired retention of the temporal component of the task (leaving the start alley within the first 5 s of a trial), which significantly improved in controls during the hours following acquisition, whereas they had no or only slight effects on retention of the discrimination component (choice of the correct alley), which did not improve over time. This retention deficit did not appear to be due to an action on acquisition, retrieval, and/or forgetting processes, or to state-dependent effects. Moreover, gamma-LGLA, CPP or AP5, when administered immediately after partial acquisition of a food-reinforced lever-press task, suppressed the spontaneous improvement in posttraining performance (PTP) observed in control mice 24 h after the training session. MCPG, an antagonist of mGluRs, also suppressed the spontaneous improvement of PTP and its effects were antagonized by the co-administration of trans-ACPD, an agonist of mGluRs.

Spontaneous improvement of PTP over time is thought to reflect an active and dynamic process, implicating organization of memory traces. According to this hypothesis, our results suggest that synaptic plasticity mediated by NMDA receptors and/or mGluRs activation is involved in mechanisms underlying long-term consolidation of memory traces.

## Symposium 13

### Sym13/1

EARLY VISUAL DEVELOPMENT OF INFANTS WITH CORTICAL AND SUBCORTICAL BRAIN LESIONS. *Giovanni Cioni. Institute of Developmental Neurology, Psychiatry and Ed.Psychology, University of Pisa and Stella Maris Scientific Institute, 56018 Calambrone, Pisa Italy*

The contribution of cortical and subcortical structures to the development of visual perception for various characteristics of the object (location, orientation, colour, size ...) has been underlined by recent models of human vision and its development. These models are mainly supported by studies carried out in normal infants and children. In the last few years, reliable, easy to perform, behavioural and electrophysiological methods of testing the visual function have been introduced in clinical practice of newborns and small infants. Moreover, non invasive neuroimaging techniques, such as cranial ultrasound (US), computed tomography (CT) and magnetic resonance imaging (MRI) are now widely used to identify *in vivo,* from the first days of life, the presence of lesions in the neonatal brain. These new diagnostic tools have been applied to infants with perinatal hypoxic-ischaemic or haemorrhagic brain insults by several authors. A high incidence of visual deficits due to lesions of posterior visual pathways has been reported in young subjects with brain lesions revealed by neuroimaging. In our Unit since 1990 several neonates at high-risk for brain damage have been submitted to longitudinal testing of various visual functions (including grating acuity, visual field size, fixation and following, OKN, VEP), neonatal cranial US followed by brain MRI. All infants have been followed in the first two years of life and some of them until the age of 5. The results of this study have indicated optic radiation, visual cortex, basal ganglia and parietal cortex as the brain structures more likely to be involved in different types of visual impairment. A relationship between side and size of brain lesions, as shown by brain MRI, and type and severity of visual impairment was found. Moreover, longitudinal assessments have shown in different infants a gradual recovery or a worsening of visual functions during the first months of life, as a consequence of timing, characteristics of brain lesion and other factors which are difficult to identify. Visual impairment clearly correlated also with motor and cognitive deficits of these patients. The data obtained on these patients might be significant for a better understanding of the involvement of subcortical and cortical structure in early visual development.

### Sym13/2

SELF-PERCEPTION AND ACTION IN INFANCY. *Philippe Rochat. Emory University. Atlanta, GA 30322, USA*

By 2–3 months, infants spend a great deal of time exploring their own body moving and acting in the environment. Invariably, young infants spend long period of time watching their own hands, arms, and legs moving in space. They babble and touch their own body, attracted and actively involved in investigating the rich intermodal redundancies, temporal contingencies, and spatial congruence of self-perception.

If early on infants are captured by the simultaneous experience of hearing themselves, seeing and feeling the limbs of their own body moving in space, the question is whether infants, beyond a mere fascination, are actually detecting the intermodal invariants specifically attached to self-produced movements. Recent research investigating the spatial and temporal determinants of self-perception early in development is presented. Based on this research and as a general theoretical framework, it is proposed that from birth, infants are actively involved in self-exploration. Based on this self-exploration, young infants rapidly develop an ability to detect intermodal invariants and regularities in their sensorimotor experience that specify themselves as separate entities, agent in the environment. In support of this proposition, recent observations on the detection of intermodal invariants regarding self-produced leg movements and auditory feedback of sucking by young infants are reported.

These observations demonstrate that early in development, and long before mirror self-recognition, infants develop a perceptual ability to specify themselves. It is tentatively proposed that young infants' propensity to engage in self-perception and systematic exploration of the perceptual consequences of their own action plays an important role in the intermodal calibration of the body and is probably at the origins of an early sense of self: the ecological self.

### Sym13/3
ABOUT FUNCTIONAL BRAIN SPECIALIZATION: THE DEVELOPMENT OF FACE RECOGNITION. *S. de Schonen\*, O. Pascalis\*, C. Deruelle\*, F. Liegeois\*, J. Mancini\*\*, M. Fabre-Grenet\*\*\*. \* Developmental Neurocognition Group, Center for Research in Cognitive Neuroscience, CNRS, 31 Chemin Joseph Aiguier, 13402 Marseille Cedex 20, France. \*\* Service de Neurop,diatrie, CHU La Timone, Marseille, France. \*\*\* Service de N,onatologie, CHU Hopital Nord, Marseille, France*

Here we address the question of how face processing abilities develop during the first year of life. There is considerable evidence for specific cortical involvement in individual face processing in adults. Does that mean that brain networks are prespecified in infancy? We shall relate (a) data coming from several studies conducted in our lab and in several other labs on neonate visual preference for the face schema, on mother's face recognition and pre-episodic memory for individual faces during the first 4 months of life, and on pattern processing, to (b) data on cortical maturation in infants as assessed by PET scan (with H2 O15) and ERP recordings during face processing tasks and with data on post-lesional functional plasticity for individual face and pattern processing in childhood. We shall attempt to tell how far we can answer old questions such as "Are there prespecified components of face processing competences? Is face processing a modular system right from its begining as it seems to be in adult?" "Are there several possible developmental paths for one competence?".

### Sym13/4
INNATE AND LEARNED PERCEPTUAL ABILITIES IN THE NEWBORN INFANT. *Alan Slater. Department of Psychology, Washington Singer Laboratories, University of Exeter, Exeter, EX4 4QG*

From research carried out in recent years it has become apparent that the visual world of the newborn baby (in the period 0–7 days from birth), is highly organized. It is also clear that the newborn infant is an extremely competent learner, and can learn about, and form associations between, visual stimuli after only a short (i.e., 1–2 minutes) exposure. Accordingly, it is often not clear whether the visual and perceptual abilities that newborn infants display are the result of innate capacities or learning. These themes are illustrated with respect to two areas of research: face perception and intermodal perception. It appears that human faces are "special" in that newborn babies respond to them as faces, they prefer to look at attractive faces, and their face perception is orientation specific. While the findings on face perception can easily be interpreted in terms of an innate facial representation or "module", other evidence suggests that newborns can form auditory-visual intermodal connections after only a short exposure to the stimulation. A model is presented suggesting that innate capacities and predispositions facilitate and direct early learning.

### Sym13/5
VISUAL PERCEPTUAL ERFORMANCE OF 5 YEAR OLD CHILDREN IN RELATION TO NEONATAL ULTRASOUND ABNORMALITIES. *E. Vandenbussche[1, 2], P. Stiers[1, 2], M. Haers[1], B. M. van den Hout[4], L. S. de Vries[5], O. van Nieuwenhuizen[4]. [1] Laboratory of Neuropsychology, K.U.Leuven, Medical School, Herestraat 49, B-3000 Leuven, Belgium. [2] Centre for Developmental Disabilities, University Hospital, Kapucijnenvoer 33, B-3000 Leuven, Belgium. [4] Department of Child Neurology, and [5] Department of Neonatology, Wilhelmina Children's Hospital, Nieuwe Gracht 137, 3512 LK Utrecht, the Netherlands*

This study investigates the occurrence of visual perceptual deficits in children after neonatal brain damage. Six visual object recognition and two visuo-constructive tasks were presented to 54 children aged 5.02 to 5.76 yr. All subjects were neonatal at risk due to prematurity or birth asphyxia. From neonatal ultrasound scans, the occurrence of intracranial hemorrhage (ICH, N = 21), periventricular leukomalacia (PVL, N = 21), and/or white matter damage (WMD, N = 14) due to either of these conditions was determined for each subject. Scans were normal in 20 of them. In order to assess the separate contributions of mental and perceptual abilities, mental as well as chronological age was used to compare the subjects' performance on each task with that of normal children. Mental age was assessed from performance intelligence data. Forty-one children were tested with the McCarthy Scales of Children's Abilities. For the remaining 13 children data from clinical records were used.
The number of subjects performing at or below Pc5 of same-age normal children (CA condition) was significantly above 5% for all but one task (range 6–33%). This was still true for one visuo-constructive and four object recognition tasks when mental instead of chronological age was used for comparison (MA condition) (range 2–36%). This high incidence of impairment is not attributable to visual acuity loss, since grating acuity was reduced in only four subjects (range 14–19 c/deg). The relationship of the frequency of scores È Pc5 with the ultrasound conditions was as follows: (1) there was no relation with ICH, neither in the CA condition, nor in the MA condition (Fisher exact test, p È .05). (2) The association with PVL was significant in five tasks in the CA condition and in four tasks in the MA condition. (3) The significant association with WMD in five tasks in the CA condition, was absent in the MA condition.
These results show that performance on visual perceptual tasks is related both to PVL and WMD, when chronological age is used as the baseline for evaluation. However, when the visual perceptual component in these tasks is singled out by using mental age as the baseline, the relation is only with PVL. This suggests, that WMD is related to mental deficits rather than impairments in the visual perceptual domain. This is confirmed by the fact that half of the WMD cases comprise of porencephalic cysts, which are extensive but unilateral – in contrast to PVL, which is bilateral. On the other hand, transient echographic densities (PVL grade I, de Vries et al., 1992, *Behav. Brian Res.*, 49: 1–6), which are not considered as WMD, are related to visual perceptual performance in the MA condition. We conclude that although WMD can give rise to general functional impairments, PVL seems to be specifically indicative of visual perceptual impairments, regardless of mental disability or visual acuity loss.

Supported by Praeventiefonds, the Netherlands, nr 2814061.

### Sym13/6
INFANT'S DEVELOPMENT OF OBJECT PERCEPTION AND RECOGNITION: BEHAVIOR AND BRAIN IMAGERY. *François Vital-Durand. Cerveau et Vision, INSERM Unité 371, 69675 BRON cedex – France*

Behavioural techniques providing precise descriptions of young infants perceptual capacities and their evolution with time and experience can be used in parallel with newly available imaging tools for assessing the maturational or lesional condition of the brain. Cranial ultrasound (US), computed tomography (CT), PET scan and magnetic resonance imaging (IMR) provide quantitative evaluation of the maturational course of the main visual processing pathways: (dorsal and ventral) and the role of a newly discovered specific visual area (lateral occipital or LO) with their hemispheric specializations. The data shown by the contributors will be helpful in defining models of development which include hemispheric, parallel and hierarchical processing. The discussion can be enlarged to include constraints imposed by the relative course of maturation of feedforward and feedback loops.

This knowledge should be used to orient the remediation programs that can be provided to infants affected by various types of lesions by sharpening the prediction of their future capacities based on instrumentation of spared pathways and vicariant processes.

## Symposium 14

### Sym14/1
REORGANIZATION OF CIRCUITRY WITHIN THE RAT BARREL FIELD FOLLOWING INNOCUOUS CHANGES IN TACTILE EXPERIENCE. *Michael Armstrong-James. Department of Physiology, Basic Medical Sciences, Queen Mary and Westfield College, London University, Mile End, London E14NS*

Following innocuous changes in somatosensory experience at maturity, modifications of thalamic, thalamocortical and intracortical synaptic relays may contribute to receptive field plasticity of S1 cortex. We have investigated these separate aspects of reorganisation in the barrel cortex and the barreloid (Vpm) thalamus of adult rats following 3 to 30 days of cutting all but two whiskers unilaterally. Statistical analysis of changes in spatio-temporal features of responses in homologous barrel and barreloid neuronal populations reveal the degree by which use-dependent modifications in the thalamic relay contribute to processing of sensory information at the cortical level. These features will be discussed in relationship to models for experience-dependent reorganisation of somatosensory cortical maps.

### Sym14/2
EXPERIENCE – DEPENDENT CHANGES IN FUNCTION AND ANATOMY OF ADULT BARREL CORTEX. *Malgorzata Kossut. Department of Neurophysiology, Nencki Institute, 3 Pasteur st., 02-093 Warsaw, Poland*

Manipulations of sensory input to vibrissal mechanoreceptors can modify columnar functioning of the barrel cortex. Mapping of functional activity of sensory cortex with [14]C-2-deoxyglucose reveals that stimulation of a single vibrissa activates a functional column, centered upon the appropriate barrel, and extending through all cortical layers. In mice, partial vibrissectomy sparing one row of vibrissae in young adults, results, 7 days later, in an increase of the functional columns activated by the spared whiskers. The increase of the extent of labeled area is visible in all cortical layers, but particularly in layer V. If the post-lesion time is about 2 months, the increase of areal extent of labeling is accentuated, particularly in layers II/III. Investigations of cortico-cortical connections between the rows of barrels were carried out in living cortical slices of the barrel cortex of mice two months after vibrissectomy sparing row C. Antero – and retrograde transport of fluorescent dextrans injected into identified vibrissal columns was analyzed with standard and confocal microscopy. Elongation of axons originating in the spared vibrissae representation was observed. The deprived column also sent out longer axons running in layers II/III than normal control one and received input from greater distances than normally at the level of layer II/III and V. Increased branching of axons connecting the spared column was observed in layers I through IV. Thus, prolonged changes of functional activation of adult barrel cortex are accompanied by rearrangement of cortico-cortical circuitry.

### Sym14/3
ADULT SENSORY PLASTICITY WITHIN THE BARREL-COLUMN OF MICE. *Egbert Welker: Institut d'Anatomie; Rue du Bugnon 9; 1005 Lausanne, Switzerland*

The anatomical definition of whisker-representations in the somatosensory pathway of mice and rats forms an important factor in the interpretation of modifications in the whisker-to-barrel pathway in studies on plasticity. It allows the identification of the part of the CNS where, upon e.g. peripheral manipulation, central modifications predominate. Here, findings within the pathway from experiments inducing plasticity in adult mice will be discussed. Two paradigms have been used: (1) sensory deprivation as a consequence of a peripheral nerve transection and conversely, (2) increased sensory stimulation selectively applied to individual whiskers. Using anatomical and physiological techniques to compare the outcomes of these two types of approach, we conclude that the neurones in cortical layer IV actively regulate the entry of sensory information into the entire cortical column. The molecular mechanisms which may underlie this regulation will be discussed.

# Disease and Pathology

## DP/1

A MICROANATOMICAL STADY OF THE HUMAN FETAL TRIGEMINAL GANGLION. *S. Arsic[1]. [1] Dept.of Anatomy, Nis University Faculty of Medicine, 18000 Nis, Yugoslavia*

As the pain caused by the trigeminal neuralgy is one of the strongest pain known in the humans, the investigations of the one gasserian ganglion and it's topographical relations with the surrounding structures are always actual. The aim of the investigation was to determine characteristics of the external arterial blood vessels of the ganglion and their relation with the surrounding structures especially with the cranial nerves. The examination was performed on the 50 fetal trigeminal ganglions. Fetuses of both sexes, age over 2 lunar months, were obtained after spontaneous or artificial abortion. The age of the fetuses was determined by measuring the crow-rump distance. After the application of the contrast material "micropaque" in a.carotis communis, a microdissection of the cavum semilunare dura mater, under the control of the surgical microscope, and an analysis of the arterial blood vessels and surrounding cranial nerves were made. The results of examinations show that trigeminal ganglion is vascularised prevalently by the branches of the a.carotis interna and a.meningea media. Dorsal meningeal(clival) artery vascularised the trunk of the n.trigeminus and n.abducens. N.mandibularis (V3) is vascularised by the branches of the a.meningea media. A.cerebellaris superior also vascularised the ganglion and the trunk of the trigeminal nerve. The study aims to contributing to a better understanding of the relations between gasserian ganglion and surrounding structures which are important for neurosurgery.

## DP/2

ABSENCE OF IMPLICIT LEARNING IN CHRONIC SCHIZOPHRENIA: EVIDENCE FROM A TASK REINFORCING SLOW RESPONDING. *P. Brugger[1], R.E. Graves[2]. [1] Neurology Clinic, University Hospital Zurich, CH-8091 Zurich, Switzerland. [2] Dept. of Psychology, University of Victoria, Box 3050, Victoria, British Columbia, Canada V8W 3P5*

We introduce a task designed for the study of the relationships between implicit and explicit learning. Subjects press four keys a self-determined number of times in any order. Termination of one trial is indicated by the subject and immediately followed by either positive or negative feedback. During 100 trials, subjects have to find out on what behavior the type of feedback depends. Unknown to subjects, the contingency is time-based; whenever a trial lasts more than 4 seconds, it is a "success", faster trials are "failures". We administered this task to patients with acute (n = 7) and chronic (n = 8) schizophrenia and to two respective control groups. Both control groups and acute patients showed implicit learning, i.e., time per trial increased with increasing task duration despite subjects' unawareness about the temporal character of the contingency. (Typically, subjects attributed successful performance to particular sequences of key presses which were more or less complex and, hence, time-consuming). Chronic patients showed "inverse" learning, i.e., due to initially slow responding, they were successful at the beginning of the task, but performance deteriorated due to a marked restriction to only a few stereotyped sequences which were pressed with steadily increasing speed. Behavioral stereotypy and lack of implicit learning are interpreted as functional "hypofrontality" and are consistent with a hypoperfusion of frontal lobe structures reportedly characteristic of chronic schizophrenia.

## DP/3

HOW AND TO WHAT EXTENT DOES THE NEUROPSYCHO-LOGICAL ASSESSMENT PREDICT BRAIN PERFUSION ABNORMALITIES IN PATIENTS WITH CEREBRAL DEGENERATIVE DISEASES? *A. Cappa[1], A. Giordano[2], M.L. Calcagni[2], G. Villa[1], G. De Rossi[2], G. Gainotti[1]. Institutes of [1] Neurology and [2] Nuclear Medicine of the Catholic University of the Sacred Heart, Largo A. Gemelli, 8 – 00168 Rome, Italy.*

The aim of our study was to correlate the neuropsychological pattern with the location of perfusion abnormalities in different degenerative cerebral diseases.

We studied 45 patients: 12 control subjects (C group), 20 patients with Alzheimer Disease (AD) divided into a "focal" AD group (6 patients with a prevalent memory deficit) and a "diffuse" AD group (14 patients with a diffuse and homogeneous cognitive impairment), 6 patients with Progressive Supranuclear Palsy (PSP group) and 7 patients with Circumscribed Cortical Degeneration (CCD group) showing a heterogeneous pattern of clinical symptoms, without or with minor memory deficits.

All patients underwent a neuropsychological assessment (including MMSE, Mental Deterioration Battery and a set of 15 neuropsychological tests in order to evaluate different cognitive abilities) and SPECT imaging with Tc99m-HM-PAO using a four heads dedicated tomograph (Certo-96). Tracer uptake was semiquantified in 29 regions of interest, including temporo-mesial areas (hippocampal-parahippocampal structures) according to a previously validated procedure. Statistical analysis was performed by the Kruskall-Wallis test with Bonferroni's correction where appropriate.

"Focal" AD group showed significant left temporo-mesial and temporo-lateral hypoperfusion as compared to C group (p = 0.003 and p = 0.006, respectively). "Diffuse" AD group showed severe hypoperfusion in several cortical regions, constantly including the left posterior temporo-parietal region (p = 0.002 vs C group). A significant correlation was found between the number of cortical regions involved and the number of neuropsychological tests with abnormal score (r = 0.78, p < 0.01).

In PSP group a significant hypoperfusion was found in left and right superior frontal regions (p = 0.003 and p = 0.002, respectively) and in right parietal region (p = 0.006).

In CCD group cortical region hypoperfusion was located in each patient according to the individual pattern of cognitive impairment.

A significant bilateral anterior temporal perfusion defect was found in 7 patients with memory impairment as compared to 6 patients without memory impairment (the grouping included only the 13 patients of PSP and CCD groups): p = 0.01.

Conclusions: (1) the typical SPECT pattern of bilateral posterior temporo-parietal perfusion defect was found only in the "diffuse" AD group while the "focal" AD group showed a constant left temporal defect; this pattern seems to be a correlate of the memory impairment rather than a sort of "biological" marker of early AD. Futher support to this inference is provided by the finding of a constant temporal hypoperfusion in non-AD patients with memory impairment. (2) A tight correlation was found between the severity of cognitive impairment and the extension of cerebral cortical hypoperfusion. (3) In PSP group and CCD group the location of perfusion defect confirmed the cerebral areas involved in the impaired neuropsychological abilities.

## DP/4

NEUROMORPHOLOGICAL AND NEUROCHEMICAL FEATURES IN TS65DN MOUSE, A MODEL FOR DOWN SYNDROME. *M. Dierssen, I.F. Vallina, C. Baamonde, C. Martinez-Cue, M. Lumbreras, J. Florez. Dept. Phisiology and Pharmacology, U. Cantabria, 39011 Santander, Spain*

Mice with segmental trisomy 16 (Ts65Dn) which have a triplication of a region of MMU-16 homologous to q22 region of HSA21 have been studied. No differences in gross morphology were detected. Immunocytochemical and hystochemical techniques did not reveal differences in diferent neuronal markers. However, at 15 months of age, stereological technique revealed a smaller volume of the hippocampal formation in trisomic mice. The analysis of several neurotransmitter system showed no alteration in different markers for the cholinergic system in the adult mice. Immunoreactivity for ChAT or ACE and binding for muscarinic receptor in the cerebral cortex presented no abnormalities in Ts65Dn. With respect to the norAdrenergic system, radioligand binding studies revealed a slight reduction in affinity in trisomics, with no differences in receptor densities. Cyclic AMP formation was determined in basal conditions and under stimulation with isoprenaline (ISO) or forskolin (FK). Basal cAMP production was significantly reduced in the hippocampus and cerebral cortex of Ts65Dn mice. Stimulation of cAMP production was less effective in trisomic mice, both after beta adrenoceptor stimulation or after direct stimulation of the catalytic subunit of adenylil cyclase in the hippocampus and cerebral cortex. Although no important differences were observed at the neuromorphological level, we conclude that Ts65Dn have severe deficiencies in synaptic transmission, that affect noradrenergic function in specific brain areas. (Supported by R. Areces and M. Botin Foundations and DGICYT PC94-1063; CMC and IFV have predoctoral fellowships from Gobierno Vasco and Real Patronato para la Prevención y Atención a Personas con Discapacidad, respectively).

**DP/5**

SPATIAL-PERCEPTUAL AND BODY-SCHEME DISTURBANCES IN CHILDREN WITH BENIGN ROLANDIC EPILEPSY AND CENTRO-TEMPORAL DISCHARGES. *M. Feucht , S. Voelkl, R. Studener, R. Goessler, F. Uhl\*. Universitätsklinik für Neuropsychiatrie des Kindes- und Jugendalters Wien, AKH, A-1090 Vienna. * Neurologische Universitätsklinik Wien, AKH, A-1090 Vienna*

Problems of learning and behaviour are overrepresented in children with epilepsy. They occur in estimated 5–15%, depending on ictal/interictal spike activity, the localization of the epileptogenic zone and the central side-effects of antiepileptic drugs (AEDs). The diagnostic evaluation of pediatric patients should therefore include objective, quantified measurement of neuropsychological functioning. We examined children with centrotemporal sharp wave foci, in order to reveal specific patterns of neuropsychological dysfunction, congruent with the clinical and EEG findings.
25 children (8 girls and 17 boys), between 6,1 and 10,4 years of age (mean 8,9) have been examined: Fifteen of them suffered from typical rolandic epilepsy, the other ten exhibited the characteristic EEG spikes without overt seizures.
Handedness (according to examination with the Edinburgh Inventory) was right in 19 cases, left in 2 cases and bilateral in 4 cases.
Intellectual abilities were assessed with the Hamburg-Wechsler Intelligence Scale for Children (WISC-R). Full-scale IQ was 94–109 (mean 103.9), performance score was 95–110 (mean 100), verbal score was 91–114 (mean 106,3).
Neuropsychological assessment of basal perceptual functions, using "The Neuropsychological Test Inventory for Basal Perceptual Functions" (G. Spiel et al.) revealed deficits in the dimension "Spatial Perception" in 22 cases (88%) and the dimension "Body Scheme" in 20 cases (80%). There were no differences between patients with and those without AED treatment.

**DP/6**

IMPAIRED INTEGRATION OF AUDITION AND VISION IN SCHIZOPHRENICS. *Beatrice de Gelder (1), Josef Parnas (2), Pierre Bovet (3), Jean Vroomen (1) and Theo Popelier (1). (1) Tilburg University, Department of Psychology, P.O.Box 90153, 5000 LE Tilburg, The Netherlands. (2) Psykologisk Institut, Kommunehospitalet, Copenhagen, Denmark. (3) Policlinique psychiatrique universitaire A, Lausanne, Switzerland*

Audition and vision are two input channels that are normally combined in a single percept as has been shown for speech by the McGurk effect (combination of lipreading and hearing, McGurk and McDonald, 1976) and similarly for expression (combination of face and voice expressions, de Gelder and Vroomen, 1995).

The goal of our study was to examine whether schizophrenics showed a problem in integrating sensory input from different modalities rather than having problems with processing in each of these modalities taken separately. A group of chronic schizophrenics was studied none of which suffered auditory hallucinations. Three audiovisual tasks were administered and comparisons were made with data from normal control subjects.
Two tasks examined the integrated perception of emotions. The results show an impaired recognition of vocal impression but more importantly, but above and beyond that they illustrate a lack of normal integration of expressive information form the face with prosodic information from the voice.
The audiovisual expression recognition tasks used a cross-modal bias paradigm. The first task examined the effect of a voice expression (happy or sad) on a face categorisation task using a morphed happy to sad face continuum. The second task studied the reverse and looked at the effect of a face expression (happy or afraid) on the judgement of a voice expression using stimuli from a morphed voice continuum (happy to fear).
Recognition of facial expressions in the unimodal condition was nearly normal but contrary to normal controls schizophrenics did not show the classical pattern of a cross-modal bias. There was no displacement of the category boundary as a function of the voice expression. Such a cross-modal bias effect was also absent in the second task.
The third task examined audiovisual integration for speech. Stimuli consisted of a videotape of spoken syllables produced by factorially combining a 5 step speech continuum (ba to da) with a 5 step lip-movement continuum (ba to da) Massaro & Cohen, 1990). There was also an integration deficit here albeit less marked than with the expression tasks.
Understanding problems of integration of information coming from different sensory modalities may be a critical for research on schizophrenia.
de Gelder, B. & Vroomen, J. (1995). Hearing smiles and seeing cries: The bimodal perception of emotion. Thirty-sixth Annual Meeting, Psychonomic Society, Los Angeles, California
de Gelder, B. & Vroomen, J. (1996) Categorical perception of emotional speech. The Journal of the Acoustical Society. 100, 4, Pt. 2, 2818
Massaro, D.W. & Cohen, M.M. (1990) Perception of synthesized audible and visible speech. Psychological Science. 1, 1–9

**DP/7**

PATIENTS WITH PARKINSON'S DISEASE AND AGE MATCHED CONTROLS SHOW THE SAME PATTERN OF PERFORMANCE IN TESTS OF VISUAL CONTRAST SENSITIVITY. *P. Goddard[1], C. Clarke[2], K. Kiernan[1]. [1] Cognitive Neuroscience Research Group, Department of Psychology, University of Lincolnshire and Humberside, Hull, UK. [2] Department of Neurology, Hull Royal Infirmary, Hull, UK*

Purpose. Parkinson's disease (PD) is associated with a drop in visual contrast sensitivity. However, there is wide variation between individuals in psychophysical tests of contrast sensitivity[3] and low sensitivity is also associated with normal aging[4]. This means that an individual's sensitivity can not be used as a pointer to a retinal dopaminergic deficit but it may be the case that a systematic change in the pattern of an individual's performance can. Our aim was to compare the pattern of performance between individuals in visual detection tasks using static and temporally modulated stimuli.
Method. Stimuli were generated using a Visual Stimulus Generator 2/1 (Cambridge Research Systems) and were presented on the screen of a Mitsubishi HL7955 colour monitor ( mean luminance 50 cd/m$^2$, frame rate 100 Hz). The monitor was gamma corrected to ensure that the voltage to luminance relationship was linear.
All stimuli were achromatic, horizontally-oriented, simple sinusoidal gratings presented within a 10 degree circular window. Three different spatial frequencies were used (0.5, 5 and 10 cycles/degree). For each spatial frequency a grating was presented either static or temporally modulated in counterphase at 4 Hz.

A self paced method of adjustment was used. The observer changed contrast until the stimulus was just visible. The contrast detection threshold recorded for each stimulus was the mean of six estimates.
Participants. Three groups were used, 1. nine patients diagnosed with idiopathic Parkinson's disease tested on medication 2. the same group of patients tested off medication and 3. an age matched control group.
Results. Individuals in each group showed the same pattern of performance. With counterphase flicker, sensitivity increased at the low spatial frequency but decreased at the medium and high frequencies relative to the sensitivities found with stationary gratings. Statistical analyses revealed no significant differences between the sensitivities of the three groups.
Discussion. We fail to show any systematic difference in the pattern of performance between patients with PD and controls. The results are not consistent with a description of the visual impairment in PD based on a selective deficit for moving stimuli, or for a reduction in sensitivity for specific narrow ranges of spatial frequency.
Within each group there was a wide variation between the sensitivities of participants. This is due partly to individual variations but is also likely to be the result of the methodology. Method of adjustment allows quick estimates of sensory thresholds but is prone to observer dependent criteria that might shroud differences in sensitivity between groups. Nevertheless, despite variations in sensitivity between observers their pattern of performance appears consistent and robust.
3. Ginsburg A P, Evans D W, Cannon Jr. M W (1984) Large-sample norms for contrast sensitivity. American Journal of Optometry and Physiological Optics, 61, 80–84
4. Mestre D, Blin O, Serratrice G & Pailhous J (1990) Spatiotemporal contrast sensitivity differs in normal aging and Parkinson's Disease. Neurology, 40, 1710–1714

## DP/8

AUTOIMMUNE REACTIONS AGAINST TROFIC FACTORS S100 IN INFANTS WITH PRENATAL BRAIN DISTURBANCES AND HYPOTROPHY. Gruden, M.A., Kurbatova, L.A., Òobolin, V. A., Shumova, E.A., Sherstnev, V.V. P.K. Anokhin Institute of normal physiology, RAMS, B. Nikitskaya,6, 103009 Moscow, Russia

Participation of trafic factors S100 in human central and peripheral functional system formation was investigated. Comparative clinical-experimental analysis of autoimmune reactions to S100 (proteins with growth and neurotrophic properties) in offspring with various forms of prenatal hypotrophy and brain alterations was carry out. Clinical, neurosonographical, immunological and immunochemical methods during investigation were explorated. In all sera samples of infants with genetic determined forms of hypotrophy ( in 27,3% accompined by marked structural changes in brain and in 94,5% – rough psichomotorical developmental delay, for example infants with Daune-syndrom) autoantibodies (a-AB) to S100 in titres in 2–4 time higher that control ones were revealed without any correlation with hypotrophy degree. Conformity between a-AB-S100 and hypotrophic degree of second form hypotophy was determined, a-AB level was higher in 10–20 time from control in 70%, 75% , 100% with I, II, III hypotrophic degrees, respectively.Thus, distinct differences in autoimmune reactions against trofic factors S100 expression in offspring with cerebral changes during firth and second prenatal hypotrophy forms were carried out. Data testified the protein S100 participation in central and peripheral sections of organism functional systems is defined not only by genetic determined mechanisms

## DP/9

ASSESSMENT OF b-AMYLOID-INDUCED BEHAVIORAL DYSFUNCTIONS IN THE SMALL OPEN-FIELD PARADIGM: A SENSITIVE SCREENING OF b-AMYLOID NEUROTOXICITY AND THE EFFICACY OF NEUROPROTECTION. Harkany, T.[1,3], Ábrahám, I.[2,3], De Jong, G.I.[3], Rensink, A.A.M.[3], Sasvári, M.[1], Nyakas, C.[1,3], Varga, J.[4], Zarándi, M.[4], Penke, B.[4], Leonard, B.E.[5] and Luiten, P.G.M.[3]. [1] CRD Haynal University of Health Sciences and [2] Institute of Experimental Medicine, Budapest, Hungary; [3] Dept. of Animal Physiology, RuG, The Netherlands; [4] Dept. of Medical Chemistry, Szent-Györgyi Albert Medical University, Szeged, Hungary and [5] Dept. of Pharmacology, UCG, Ireland

One of the specific neuropathological features of Alzheimer's disease is the enhanced generation and subsequent accumulation of b-amyloid peptides (Ab). Ample evidence indicates that Ab exert cholinotoxicity and might thereby contribute to the exacerbation of learning and memory deficits in the disease process. Although several attempts were made to characterize the time of onset, progression and severity of behavioral impairments elicited by exogeneous Ab administration, by far no rapid and reliable screening task was found to prove the efficacy of neuroprotective drug treatments. In the present experiments we investigated by means of testing the animals in the small open-field task and acetylcholinesterase (AChE) histochemistry the time courses of alterations in spontaneous animal behavior and their correlation with the loss of cortical cholinergic innervation after unilateral $Ab_{(1-42)}$ infusion into the magnocellular nucleus basalis (MBN) of young and aged rats. Furthermore, the neuroprotective potentials of NMDA receptor-blockade and chronic corticosterone administration were aslo determined.
Young adult and aged male Wistar rats were injected with $Ab_{(1-42)}$ (0.2 nmol/1 ml) into the right MBN. Alterations in spontaneous behaviors were determined in the small open-field 3, 7 and 14 days post-surgery. MK-801 was administered acutely in a dose of 5 mg/kg body weight (i.p.) 2 hours prior to $Ab_{(1-42)}$ infusion. In another studies, by using a 7-day post-infusion period, corticosterone pellets (100%) were implanted (sc.) 7 days prior to $Ab_{(1-42)}$ infusion. Following the appropriate survival time cortical AChE-positive fibers were visualized according to Hedreen et al. (1985) and their loss was quantified in the posterior somatosensory cortex.
$Ab_{(1-42)}$ elicited a dramatic depression of rearing and ambulation activities in the small open-field paradigm as early as 3 days post-surgery which persisted throughout the survival periods. In parallel with altered animal behavior a ~16–21% loss of cortical cholinergic fibers was recorded. Interestingly, the extent of cortical fiber reduction and the number of explorative rearing in the behavioral test showed a strong correlation (corr. coeff.: 0.92). Ab infusion in senescent animals did not result in enhanced neurotoxicity.
Both MK-801 and corticosterone protected against Ab neurotoxicity. Drug-treated animals became transiently hyperactive but maintained normal explorative and motor activities. Histochemical analysis of cortical sections revealed that these drug treatments resulted in an almost complete preservation of AChE-positive innervation.
In conclusion, our studies demonstrate that Ab injections result in graded alterations in spontaneous animal behavior and age-mediated down-regulation of receptor expression may account for the decreased neurotoxicity in senescent rats.

## DP/10

Time-course of recovery of auditory attention following closed head injurY. I. Keller. Neurological Clinic Bad Aibling, 83043 Bad Aibling, Germany

According to clinical experience, a frequent consequence of closed head injury (CHI) is an impairment of attention. Manifestations of this impairment consist in a reduced speed of information processing as well as deficits of divided, selective and sustained attention. In a previous study, Keller et al. (1995) demonstrated a higher error-rate in an auditory attention task with patients 3–6 months after severe closed head injury. In the present study we have investigated the time-course of recovery of auditory attention in CHI patients 3–5 days after regaining consciousness.
We used an experimental paradigm that consisted of four subtests which comprised strings of auditory digits heard either diotically or dichotically, at either fast or slow presentation rates. Omission and commission errors were scored for each subtest and combined by an index of errors. Twelve patients with severe closed head injury were tested 3–5 weeks after closed head injury and then twice a

week for 12 weeks. The results indicate that CHI patients are initially impaired in all stimulus conditions. The time-course showed a recovery of attention to slow diotically presented stimuli within 2 weeks for 10 patients. Eight patients were able to react to fast diotically and slow dicotically presented stimuli within 7 weeks. Only 4 patients could perform in the fast dicotical condition within 12 weeks. The time-course of recovery of different attentional functions correlated well with the recovery of other cognitive abilities (e.g. orientation to time, memory).

The results indicate that multiple small lesions and diffuse axonal injuries lead to a severe impairment of auditory attention. It is proposed that the reduced access to redundant pathways to the neural knowledge net in CHI patients leads to a slowing of information processing speed. With recovery of speed of information processing other attentional functions such as selective attention as well as memory and intellectual abilities are restored.

Keller, I., Schlenker, A. and Pigache, R. (1995) Selective impairment of auditory attention following closed head injury or right cerebrovascular accidents. Cognitive Brain Research, 3, 9–15

## DP/11

MOTOR IMPAIRMENT IN PARKINSON'S DISEASE IS ASSOCIATED WITH DEFICITS IN VERBAL FLUENCY. *J. Kessler[1], H. Karbe[2], E. Kalbe[1], M. Bley[2] & W.-D. Heiss[1,2]. [1] Max-Planck-Institut für neurologische Forschung, Gleueler Str. 50, D-50931 Köln (Germany). [2] Neurologische Universitätsklinik, Joseph-Stelzmann-Str. 9, D-50924 Köln (Germany)*

*Objective:* Verbal fluency tasks performance of non-demented patients with Parkinson's disease (PD), compared to a healthy control group, has been described as either generally impaired, unimpaired, or impaired in semantic but not in phonological fluency tasks. We examined if these inconsistent results might be due to heterogeneous grouping. The PD group was splitted according to Hoehn-Yahr (H-Y) scores and to memory scores, following a performance of various generative and confrontation naming tasks.

*Patients and Methods:* 18 non-demented patients with idiopathic PD (mean age 55.4; SD = 11.8) and 24 healthy individuals (mean age 54.8; SD = 9.5) were enrolled in the study. The PD patients were graded by the H-Y scale (median: 2) and their cognitive status was assessed by a neuropsychological test battery. None of the patients was demented (e.g. Mini-Mental-Status-Test: mean 27.82; SD = 1.23). The groups were comparable in education and profession. The following generative and confrontation naming tasks were used:

– Verbal fluency (1 min/task): animals, parts of the body, professions, furniture, fruits/vegetables, articles of clothing, physical activities (verbs), positive attributes, negative attributes, surnames, cities, famous people, letters A, F, and S.
– Naming (15 items/category): animals, parts of the body, professions, furniture, fruits/vegetables, articles of clothing, physical activities.

*Results:* The PD patients did not differ significantly from the control group in letter fluency tasks, but showed significantly reduced word rate in 9 of 11 categories (p < 0.05) and still in 4 categories after - adjustment (p < 0.001). In the confrontation naming task a ceiling effect was observed in both groups. When the PD group was splitted in patients with the average performance of more than 6 words or less than 6 words in verbal selective reminding task (10 words, 5 trials), no significant differences in verbal fluency tasks could be established. A categorization with the H-Y scale ( 1.5, n = 8; >1.5, n = 10) showed a superiority of the PD patients with low H-Y scores in 4 fluency tasks (p < 0.05). Compared to the control group, PD patients with the more severe Parkinson symptomatology are significantly impaired in most verbal fluency tasks, although they show similar neuropsychological test performance compared to the PD patients with H-Y 1.5. No such significant difference was observed between the control group and PD patients with lower H-Y scores.

*Conclusion:* Verbal fluency deficits in Parkinson patients are closely related to the severity of motor deficits and not to different neuropsychological test performance. Semantic fluency is more impaired than phonological fluency. The higher bradykinesia and rigidity in this group covariates with tasks that demand cognitive flexibility and executive function.

## DP/12

MULTIMODAL SENSOMOTORIC STIMULATION IN NEUROSURGICAL REHABILITATION. *Lippert-Grüner, M., Terhaag, D. Neurosurgical Clinic, University of Cologne, Joseph-Stelzmann-Straße 9, 50931 Köln, (Direktor: Univ.-Prof. Dr. N. Klug)*

Immediate and systematic application of adequate rehabilitation are important factors for restitution of impaired brain function succeeding severe head trauma. By local integration into the primary clinic complications can be managed without delay; dangerous and cost-intensive transportation can be avoided. Local integration makes it possible to start rehabilitation therapy directly after the patient has weaned from mechanical ventilation. In our neurosurgical intensive care unit, 4 out of 16 beds are intended for early rehabilitation after severe head injury. Physcial therapists, ergotherapists, psychologists, speech therapists, art therapists and additional physicians with a specialisation in rehabilitation do support the nurses in patient care. After weaning from mechanical ventilation, therapy was performed for 4–5 hours per day. In the last years multimodal early-onset stimulation therapy is more and more playing an important role in the early rehabilitation concept. Here we report on the stimulation in comatose patients (n = 15, GCS < 9) from the neurosurgical intensive care unit. Ihe idea is to contact the comatose patient by multimodal (orofacial, accoustic, tactile, cinesthetic and proprioreceptive) stimulation. Stimulation is performed in two sets of one hour per day. During each of these sets heartrate, breathing frequency, eye movements and hydrogalvanic skin resistance of the patient is monitored continuously and behavioral changes are documented. We observed significant changes of the heartrate and skin resistance during tactile, cinesthetic and accoustic stimulation. After orofacial, accustic and visual stimulation significant behavioural changes (eye-, head- and limb movements, vocalisation, mimic responses) of the patients were observed. No significant reactions were observed after tactile, proprioceptive and cinestetic stimulation. Follow up studies will reveal if early-onset stimulation has an influence on recovery and outcome in patients with severe head injury.

## DP/13

METABOLIC AND NEUROPSYCHOLOGICAL CHANGES IN EARLY-ONSET-CEREBELLAR-ATAXIA DEMONSTRATE CEREBELLAR CONTRIBUTION TO HIGHER CORTICAL FUNCTIONS. *R. Mielke, J. Kessler, R. Hilker, G. Weber-Luxenburger, W.-D. Heiss. Max-Planck-Institut für neurologische Forschung und Universitätsklinik für Neurologie, Gleueler Str. 50, D-50931 Köln, Germany*

Recent research reports suggest that the computational properties of the cerebellum contribute to motor skills and to the modulation of higher cognitive functions as well.

We studied a family of which two members fulfilled the clinical criteria of early-onset-cerebellar-ataxia (EOCA) with retained reflexes. While their motor deficits remained stable in the last time we were able to demonstrate progressive deficits in intelligence and visuospatial abilities. Information processing speed was slowed and both patients were severely impaired on a test of verbal fluency. MRI-scans revealed a marked cerebellar parenchymal loss in our patients, while supratentorial structures showed only least sulcal prominence. Positron emission tomography (PET) with [11]C-flumazenil was used to study gamma-aminobutyric type A/benzodiazepine receptor binding (BZR), and [18]F-2-fluoro-2-deoxy-D-glucose to analyze longitudinal regional cerebral glucose metabolism.

Flumazenil-PET demonstrated a decreased BZR density in the cerebellum with predominance of the nucleus dentatus and vermis as evidence of distinct cerebellar neuronal loss. In comparison to age-matched controls, patients showed a global metabolic decline and predominant hypometabolism in the thalamus and cerebellum.

The predominant thalamic hypometabolism may be a consequence of secondary degeneration of cerebellar fibers that project via the ventroanterior and ventrolateral thalamus to motor and premotor areas, and to the parietal association cortex. Progressive cognitive decline and metabolic derangement in cortical regions and association areas may be explainable by a disturbed integrity of cognition-related networks due to secondary degeneration of cerebello-thalamo-cortical projections.

## DP/14

DEPRESSIVE STATUS AFTER NEURAL TRANSPLANTATION IN AN EXPERIMENTAL MODEL OF ALZHEIMER'S DISEASE. *N. Popovic[1], K. Jovanova-Nesic[1], M. Popovic[1], D. Bokonjic[2] and Lj. Rakic[3]. [1] Immunology Research Center "Branislav Jankovic", Vojvode Stepe 458; [2] Medical Department, Military Technical Institute, Kataniceva 15 and [3] Department of Biochemistry, School of Medicine, Pasterova 2, 11000 Belgrade, FR Yugoslavia*

Disturbances of different types of affective reactions in Alzheimer's disease (AD) have became a subject of growing number of clinical and experimental investigations. The present study was undertaken to 1) assess the depressive status of rats with the lesions of nucleus basalis magnocellularis (NBM) (an experimental model of AD) and 2) to establish the possibility for its modulation by "early" (EG) or "delayed" (DG) intracerebral neural grafting. Therefore, small fragments of fetal frontal cortex tissue (day 18 of gestation) were allo-transplanted into the lesioned NBM of adult male Wistar rats two (EG) or ten (DG) days after bilateral electrolytic lesions. Ten days later test of learned helplessness was performed, in intact (IC), sham-operated (SO), NBM-lesioned, EG- and DG-transplanted as well as sham-transplanted rats. The lesions of NBM induced significantly reduction of the latency period in preshocked animals (p < 0.001 compared to IC and SO). In comparison with NBM-lesioned and EG sham-transplanted rats, the EG-transplanted rats showed marked (p < 0.001) increase of the number of shocks delivered before the animal made the correct escape response. On the other hand, in group of DG transplanted rats the significant improvment of depressive status was not found. These results indicate that neural transplantation performed within precise defined postlesioning period can ameliorate affective reaction in NBM-lesioned rats.

## DP/15

SEIZURE ONSET PREDICTION IN CASE OF A 17-YEARS OLD GIRL. *N. Rajsic[1], S. Suljagic[2], Z. Bozovic[3], J. Ivanus[3], D. Rapaic[2], G. Nedovic[2], A. Kalauzi[3]. [1] Institute for Mental Health, Palmoti'eva 37, 11000 Belgrade, Yugoslavia. [2] Faculty of Defectology, Visokog Stevana 2, 11000 Belgrade, Yugoslavia. [3] Center for Multidisciplinary Studies, Kneza Viseslava 1, 11000 Belgrade, Yugoslavia*

The purpose of this study is to extract reliable parameters of preictal EEG activity from a patient with epilepsy that can predict the seizure. The hypothesis was that coefficient of relative delta activity CRDA = PDELTA/(PTOTAL – PDELTA) can successfully predict seizure onset. CRDA greater than 3 was adopted as predictive.
Methods: We analyzed visually and mathematically a total of 14 seizures obtained from a 17-years old female during 48 hours video and cassette EEG recording. The onsets of every seizure were identical each other clinically and electrophysologically. A total of 200 secondes, eight channels artifacts free EEG records, separated in fifty 4-seconds epochs, obtained with parasagital montage, were analyzed off-line using Discrete FFT. CRDA coefficients, for 2 to 3.75 Hz EEG delta band end for every epoch, were calculated. We summarized predictive CRDA values for every channel and analyzed its course following subsequent epoches of recording.
Results: Interical epileptiform focal transients occurred as phase reversion of small spikes in channels 1 and 2 (F4 electrode position).

According to the video records there appeared partial seizures with motor symptoms (tonic deviation of eyes to left, propulsion and trunk version to right side and rotation of whole body towards left side). Last five seizures were secondary generalized to GTCS. The mean duration of first nine seizures was 11 seconds, but of the last five was about 72 seconds.
The total number of predictive CRDA values for channels 1 to 8 respectively were: 60. 51, 105, 84, 53, 85, 114 and 90. Predictive CRDA values preponderate on channels 3 and 7 (C4-P4 and C3-P3) although focal epileptiform transients were registered under F4 location. Also, it was shown that in channels 3 and 7 at least three subsequent predictive CRDA intervals occurred 12, 56 and (especially) 160 seconds before seizure onset.
Conclusions: Our results suggest that the method described might predict seizure onset early enough.

## DP/16

NAMING AND MEMORY EXAMINATION IN THE DIFFERENTIAL DIAGNOSIS OF ALZHEIMER'S DISEASE: *B. Romero, T. Theml, I. Goetz, C. Steinmann: Psychiatrische Klinik der TU München, Ismaninger Str. 22, 81675 München*

We examined whether performance on (1) a memory and (2) a naming test in elderly depressive patients differed from that of patients with very mild Alzheimer's disease and whether such differences may aid the differential diagnosis between these conditions.
Three groups (n = 21 each), matched for age and education, were examined: Patients with very mild Alzheimer's disease (MMSE ° 25), depressive patients with the diagnosis of major depression or bipolar depressive disorder (DSM-III-R), and healthy controls.
Instruments: (1) California Verbal Learning Test (CVLT), German experimental edition (Ilmberger); as a measure for memory performance the number of words named in the delayed cued recall was chosen; (2) The naming-subtest of the Aachener Aphasie Test (AAT). Naming performance was examined only in the two groups of patients.
Results: The non-parametric data analyses showed significant group differences in recall, and significant group differences in naming performance. A more thorough analyses of the distribution of the data revealed memory deficits in a considerable part of the depressive group, whose results were overlapping with those of patients with Alzheimer's disease. Naming performance was impaired in 38% of the patients with very mild Alzheimer's disease, and in none of the depressive group.
Conclusions:
(1) The overlap we found between the memory performance in the group with a diagnosis of depression and that in the group of Alzheimer's disease indicates the limited usefulness of this kind of memory examination
(2) whereas the specificity of the naming task underlines its relevance for the differential diagnosis between Alzheimer's disease and depression in individual cases,
(3) In 38% of our group with Alzheimer's disease, in which the dementia syndrome could not be shown by the MMSE (MMSE ° 25), deficits were found in the naming task. This demonstrates the potential relevance of a language examination in the early recognition of Alzheimer's disease.

## DP/17

DEFICIT HYPERACTIVITY DOSORDER SEEMS TO BE A REINFORCEMENT DEFICIT DISORDER – A COMPARATIVE STUDY. *Terje Sagvolden[1], Heidi Aase[1], David F. Berger[2], and Pål Zeiner[3]. [1] Department of Neurophysiology, University of Oslo, P.O. Box 1104 Blindern, N0317-Oslo, Norway; [2] Department of Psychology, State University College at Cortland, Cortland, N.Y., USA; [3] Norwegian State Centre of Child and Adolescent Psychiatry, University of Oslo, Oslo, Norway*

Attention Deficit Disorder (childhood hyperactivity) is a behaviour disorder affecting 2–5% of grade-school children. There is no brain damage in ADD, but a genetic factor gives rise to neurochemical imbalances that cause the behavioural problems: deficits in sustained attention (inability to maintain an attentional set), overactivity (inappropriate motor activity), impulsiveness, and an increased variability in all behaviour. Impulsiveness is increasingly seen as a key characteristic of the disorder. In the clinical study, we used 8 boys with Attention Deficit Hyperactivity Disorder (ADHD) and 12 control boys, all aged 7 to 12 years, to investigate the hypothesis that there are altered delay-of-reinforcement gradients in ADHD. In the animal study, we used male and female spontaneously hypertensive rats (SHR), an animal model of Attention Deficit Hyperactivity Disorder (ADHD) and male and female WKY comparison rats, 8 rats in each group. Both children' and model's behaviour was trained and maintained in lever-press response apparatuses operating according to a multiple fixed interval extinction schedule of reinforcement. Trinkets were used as reinforcers for the childrens' behaviour and drops of water for the animals' behaviour. The results showed strikingly similar behaviour in ADHD subjects and SHR rats: the hyperactive behaviour was acquired, with high frequency responding during both schedule components, development of short interresponse times (response bursts, "impulsiveness") and a sustained-attention deficit in the extinction component. Both control children and the control group ceased responding in extinction and did not develop response bursts. The behavioural characteristics of the ADHD boys and the male SHRs may be due to a combination of impaired sustained attention and altered reinforcement mechanisms. The results strengthen the position that SHR is a useful model of ADHD which can be used for investigating neurobiological and other factors underlying ADHD. Factors which cannot be investigated in humans.

## DP/18
LEARNING DISABILITIES IN A GIRL WITH EPILEPSY. *M.G. Vukovic[1], R.D. Sujic[2], D.O. Vranjes[2]. [1] Faculty of Defectology, University of Belgrade, Visokog Stevana 2, 11000 Beograd, Yuogoslavia, [2] Institute of Neurology, Zvezdara Clinical Center, Dimitrija Tucovica 161, 11000 Beograd, Yugoslavia*

A case history of a 11-year-old girl with epilepsy resulting from a meningitis in early childhood is presented. Details of neuropsychological assessment are discussed, highlighting the difficulties of learning abilities and also presented neurophusiological studies of frequent subclinical epileptiform discharges. In examination of cognitive and language functions we used Rey Osterrieth complex figure test, Auditory Verbal Learning test, Boston naming test, REVISC, Trail making test, Raven's Coloured Progressive Matrices. Neuro-

psychological findings reveal mild dicrease of intellectual capability, noticable fluctuated attention, visuoconstructive disorders, lexical-semantic deficiencies, dyslexia and dysgraphia. Anticonvulsant drugs and neuropsychological treatment, both of them improved the patient's cognitive condition.

## DP/19
THE MUELLER-LYER ILLUSION IN DEMENTED PATIENTS. *B. Weber, L. Froelich, N. Helbing, D. Simminger, K. Maurer. Department of Psychiatry and Psychotherapy I, University Hospital Frankfurt/Main, Heinrich-Hoffmann-Straße 10, D-60528 Frankfurt/Main, Germany*

The influence of perception disturbance on cognitive capabilities is evident in older people. Changes of gestalt perception and their possible influence on cognitive capability are not yet sufficiently examined. Optical gestalt perception must be regarded as an impotent organizing factor, facilitating orientation and comprehension in complex situations and tasks. Assuming a particular psychological function of optical gestalt perception we would expect a decreasing extend of optical illusion in the case of an impairment of this function. An increase of optical illusion would be expected in the case of preserved optical gestalt perception and loss of adaptability and cognitive compensation, usually revising the phenomenon of optical illusion. In our study we compared 16 demented patients to 13 schizophrenics (DSM-III-R) by a self developed computerized test of optical gestalt perception. It measures the extent of optical illusion by patient's assessment of 12 variations of the figure of Mueller-Lyer, differing in baseline length, in a randomised order. In addition we performed the Alzheimer Disease Assessment Scale (ADAS), the Syndrom Kurztest (SKT), the Mini Mental State Examination (MMST) and – in order to exclude sensoric lesions – an examination of sight and field of vision. The results of our study showed a significant increase of optical illusion in demented patients compared to the schizophrenic controls (p = 0.012). Taking into account that the extent of optical illusion by the figure of Mueller-Lyer usually is decreasing with age and was found to be increased in schizophrenics, our results contradict the hypothesis of an early restriction of optical gestalt perception in demented patients. It supports the assumption of a predominating loss of adaptability and cognitive compensation in these patients. Whether this loss precedes or covers up a possible impairment of gestalt perception can not be decided by our findings. Further studies are necessary in order to resolve this question and to evaluate the usability of our test for the early diagnosis of dementia.

## Emotion and Stress

### ES/1

DIVERGENT GENETIC SELECTION OF QUAIL LINES ALLOWS ELUCIDATION OF THE ROLE OF THE ARCHISTRIATUM IN FEAR BEHAVIOUR. *D.C. Davies[1], A.D. Mills[2], M. Hamilton[3], J.M. Faure[2], M.I. Gonzalez[4]. [1] Departments of Anatomy and Developmental Biology and [4] Obstetrics and Gynaecology, St. George's Hospital Medical School, Cranmer Terrace, Tooting, London SW17 ORE, UK. [2] Station de Recherches Avicoles, Institute National de la Recherche Agronomique, Centre de Tours – Nouzilly, 37380 Nouzilly, France. [3] Southwestern University at Georgetown Texas, 1001 East University Avenue, Georgetown, Texas 78626, USA*

The avian archistriatum has been implicated in fear behaviour. However, most tests of fear also involve social separation and it has proved difficult to distinguish between fear and social reinstatement (SR) behaviour. The duration of the tonic immobility (TI) response is a robust and relatively unambiguous measure of fear in domestic birds and treadmill behaviour is a reliable measure of SR. In quail, selection for TI responses weighted for independence from SR behaviour and *vice versa*, reveals that the expression of TI and SR can be separated at the genetic level. Thus, divergent selection for TI responses and SR behaviour provides a powerful tool for investigating the neural basis of fear and SR behaviour. Day-old, straight run offspring (F19 generation) of quail lines selected for long (LTI) or short duration of TI, or high or low SR behaviour, were anaesthetised and given bilateral electrolytic archistriatal lesions (ARCH) or sham-operation (SHAM). A third group of chicks was untreated (U). On day 3 post-hatch, the chicks were given open-field and hole-in-the-wall box timidity tests and on day 8 post-hatch they underwent a treadmill test for SR and a TI test. ARCH chicks were then killed and their brains processed for lesion site verification. In the LTI line only, ARCH chicks showed significantly shorter (@ 60%) TI durations than SHAM + U chicks ($F_{1,192} = 5.59$; $p = 0.01$). Across lines, latency to head movement (a precursor to TI termination) was significantly shorter (@ 50%) in ARCH than in SHAM + U chicks ($F_{1,49} = 7.69$; $p = 0.02$). Thus, in LTI chicks at least, the archistriatum can prolong the duration of TI. Since fear increases the duration of TI, archistriatal lesions appear to reduce the antecedent state of fear. There were no effects of archistriatal lesions in the treadmill test and therefore, the archistriatum does not appear to play a part in the expression of SR behaviour. ARCH chicks showed significantly altered behaviour in the open-field compared to SHAM + U of chicks of all lines (Wilk's Lambda = 0.87; $F_{12,181} = 2.22$; $p = 0.01$). They exhibited a number of behavioural changes that could either be indicative of reduced fear or increased SR behaviour. However, since archistriatal lesions did not affect the SR behaviour of any chicks in the treadmill test, their effects can now clearly be ascribed to reduced fear. LTI ARCH chicks showed a significantly shorter latency to 'peep door shut' in the timidity test than SHAM + U chicks of the same line ($F_{1,49} = 4.9$; $p < 0.05$). This altered vocal behaviour can now also be attributed to a reduction in fear for the reason given above. The results of these experiments demonstrate that the archistriatum plays a direct role in fear behaviour but not in sociality. Furthermore, divergent selection for fear has been shown to have a discrete neural correlate, a finding that has important implications for understanding the evolution of brain and behaviour.

### ES/2

SWIM STRESS TRIGGERS THE RELEASE OF VASOPRESSIN WITHIN THE SEPTUM IN MALE RATS. *K. Ebner, C. T. Wotjak, R. Landgraf, M. Engelmann. Max-Planck-Institut für Psychiatrie, Kraepelinstr. 2, D-80804 München, BR Deutschland*

Since several years the septum has been suggested act as an interface between emotional and higher cognitive functioning in rats. However, surprisingly little is known about chemical messengers that contribute to the cross-linkage of these capacities. In this context the neuropeptide vasopressin (AVP) might be one of the septal neurotransmitters/neuromodulators that might be involved in the generation of an appropriate behavioral response to emotionally and physically challenging situations. Indeed we recently demonstrated that application of an AVP V1 receptor antagonist into the septum resulted in an axiolytic-like effect as measured on the elevated plus-maze whereas treatment with synthetic AVP failed to alter animals' behavior. On the basis of these observations, it was hypothesized that exposure of the animal to emotionally challenging situations (such as to the plus-maze apparatus) might stimulate endogenous AVP release which cannot further be amplified by the exogenous peptide. To test this hypothesis we employed in the present study chronically implanted microdialysis probes to monitor intraseptal release of AVP before, during and after forced swimming as a combined physical/emotional stressor known to trigger intrahypothalamic release of the peptide. Swim stress caused a significant increase of AVP release in the medio-lateral (250%, $p < 0.05$) but not in the ventral septum indicating that emotionally and physically challenging situations are capable of stimulating vasopressinergic neurotransmission in a area-specific manner within the septum. Our findings provide not only additional support for the hypothesis mentioned above but also raise the question as to the physiological and behavioral impact of intraseptally released AVP. In the light of the results of various studies implying a causal involvement of septal AVP in learning and memory processes it might be speculated whether this neuropeptide represents the chemical component of the missing link between emotionality on the one hand and learning and memory on the other.

### ES/3

NORMAL PERSONALITY TRAITS RELATED TO DOPAMINE 3 AND SEROTONIN 2A RECEPTOR GENES? *T. Füreder[§], H.N. Aschauer[*], T. Stompe[*], K. Fuchs[§§], E. Gerhard[§§], W. Sieghart[§§], K. Meszaros[*], K. Hornik[§], S. Kasper[*]. [§] Institut für Statistik und Wahrscheinlichkeitstheorie, Technische Universität Wien, A-1040 Vienna, Austria. [*] University Hospital for Psychiatry, Department of General Psychiatry, A-1090 Vienna, Austria. [§§] University Hospital for Psychiatry, Department of Biochemical Psychiatry, A-1090 Vienna, Austria*

*Introduction*: Behavioral genetic studies suggest a substantial genetic basis of personality, proposing heritability rates of 30–60% for some traits (Cloninger et al., 1996). C.R. Cloninger proposed a method for clinical description and classification of personality variants (Cloninger, 1987) via a self-reporting Tridimensional Personality Questionnaire (TPQ; Cloninger, 1987). He postulated that normal personality traits can be described by variation along four dimensions. He called the four dimension Novelty Seeking (NS), Harm Avoidance (HA), Reward Dependence (RD), and Persistence (PS) hypothesizing they are mediated by the dopamine, serotonin and norepinephrine system, respectively.
*Methods & Results*: We administered the TPQ to 58 normal volunteers giving informed consent (41 females, 17 males, mean age 31.0, standard deviation 7.6). We tested a polymorphism of DRD3, a diallelic Bal I polymorphic site in the first exon located on 3q and a diallelic polymorphism of 5HT2A, a T/C substitution at position 102 in the gene located on 13q.
Based on Cloninger's theory, we had a prior hypothesis about an association of NS scores with DRD3 polymorphisms and HA scores with 5HT2A polymorphisms.
First, two $c^2$-tests, which compare expected and observed frequencies of persons in the respective extreme quartiles of TPQ scores, showed significant differences between marker 5HT2A and HA ($p = 0.05$) and between marker DRD3 and PS ($p = 0.02$).
Then we investigated whether the four dimensions of the TPQ depend on four selected variables (DRD3, 5HT2A, sex, and age). Multiple regression analysis showed that RD depends significantly on these variables ($p = 0.0027$).
Finally, stepwise regression analysis showed RD is mainly influenced by sex ($p = 0.0014$) and 5HT2A ($p = 0.05$).

*Discussion*: Our results of possible association between DRD3 and 5HT2A polymorphisms and PS and HA, respectively, and between a linear combination of 5HT2A, DRD3, age, and sex and RD are to be seen in context with the recent reports associating DRD4 receptor genotypes to NS (Ebstein et al., 1996; Benjamin et al., 1996). As with any newly proposed association our results need to be replicated before firm conclusions can be drawn. Associations between molecular biological data and personality traits could be present, but other factors obviously are influencing relationships and must be included in analyses.

REFERENCES
Benjamin, J., Li., L., Patterson, C., et al. (1996). *Nature Genetics*, **12**, 81–84.
Cloninger, C. (1987). *Arch. Gen. Psychiatry*, **44**, 573–588.
Cloninger, C., Adolfson, R., & Svrakic, N. (1996). *Nature Genetics*, **12**, 3–4.
Ebstein, R., Novick, O., Umansky, R., et al. (1996). *Nature Genetics*, **12**, 78–80.

**ES/4**

MILD STRESSORS INCREASE SEROTONIN RELEASE IN THE CEREBRAL CORTEX OF ROMAN HIGH-AVOIDANCE, BUT NOT LOW-AVOIDANCE RATS. *O. Giorgi, M.G. Corda, D. Lecca, G. Piras, and G. Di Chiara. Department of Toxicology, University of Cagliari. Viale A. Diaz 182, 09126 Cagliari, ITALY*

Roman High-Avoidance (RHA) and Roman Low-Avoidance (RLA) rats are selected and bred for respectively rapid versus poor acquisition of two-way avoidance behavior. This selection process has led to many other behavioral differences related to emotional factors, RLA rats being emotionally more reactive. To delineate further the functional factors underlying the different emotivity levels of RHA and RLA rats, we used brain microdialysis to compare the effects of a mild stressor, like tail pinch (TP, 40 min), and subconvulsant, anxiogenic doses of pentylenetetrazol (PTZ, 10 mg/kg, i.p.) on the release of serotonin (5-HT) in the prefrontal cortex (PFCx) and fronto-parietal cortex (FPCx) in both rat lines. No line-related differences were observed in the basal 5-HT output (fmol/20 μl), either in the PFCx (RHA: 5.7 ± 0.9, RLA: 6.8 ± 0.9; n = 26) or in the FPCx (RHA: 13.9 ± 1.4, RLA: 13.1 ± 1.2; n = 20). TP caused a significant increase in 5-HT release in the PFCx of RHA rats (+ 45% above the basal value 20 min after beginning of TP) but not in RLA rats. Likewise, PTZ increased significantly 5-HT release only in the PFCx of RHA rats. Similar results were obtained in the FPCx. Together, these results demonstrate that the activation of 5-HTergic pathways induced by TP and PTZ is more pronounced in the line of rats with lower emotional reactivity. Therefore, it may be proposed that the activation of the cortical 5-HTergic pathways may reflect an increased attention of the animal and/or the activation of cognitive mechanisms in an attempt to actively cope with the stressor. These mechanisms appear to be more readily triggered in "active" copers, like RHA rats than in "passive" copers, such as RLA rats.

**ES/5**

ULTRASOUND VOCALIZATION IN RATS DURING CLASSICAL AVERSIVE CONDITIONING. *P. Jelen, J. Zagrodzka, S. Soltysik[1]. Nencki Institute, 3 Pasteur Str., 02-093 Warsaw, Poland. [1] University of California Los Angeles, Neuropsychiatric Institute, Mental Retardation Research Center, Los Angeles, CA 90024, USA*

The ultrasound vocalization (USV) in rats is often used as one of measures of emotional and motivational states in many experimental procedures. It can be observed for example in sexual behavior, handling or agonistic encounters. Vocalization might be especially useful to assess the fear or anxiety level in stressful situations. It is not surprising because the components of mammalian vocalization circuit overlaps the structures engaged in production of fear and anxiety states associated with pain (rhinencephalic-diencephalic structures). The effectiveness of anxiolytic drugs in suppression of USV in stressful situations supports this point of view. In our experiments we analyzed the USV in Pavlovian aversive conditioning paradigm. The experiments were performed on rats restrained in special apparatus enabling the animals to run on the treadmill. Tailshock (3 mA, 100 ms) was used as an unconditioned stimulus (US). The effects of the following factors were tested: modality of the CS (5s light or tone), the presence and modality of conditioned inhibitor (CI; 3s tone or light respectively) that overlapped the last 3s of CS, administration (i.p.) of anxiolytic and anxiogenic drugs (diazepam – DZ and pentylenetetrazol – PTZ).
We found that the mean pattern of USV (% USV as a function of time) was similar in tone-CS and light-CS groups. In excitatory trials it consisted of tonic USV ceased during the presentation of the CS and of the reappearance of USV after tailshock. In inhibitory trials (scheduled randomly with the excitatory ones) the CI, independently of the modality, induced the partial reappearance of USV. The offset of stimuli (CS and CI) produced further increase in %USV. Although the course of USV changes was similar for both groups, the intensity was different. In the tone-CS group rats vocalized more before CS and during CI presentation than in light-CS group. The anxiolytic and anxiogenic drugs (DZ – 10 mg/kg, PTZ – 10 mg/kg) induced respectively decrease and increase in %USV.
The obtained results seem to be surprising, because one would anticipate rather opposite effect of CS as well as of CI. We conclude that in some situations the vocalization might not be the proper indicator of anxiety level per se. This measure can be masked by coping strategy in particular conditions (silent freezing for example) since in some circumstances, when rat is faced to the predator and tonic immobility becomes the only appropriate action, the USV would not be adaptive (Shepherd et al., 1992). In experimental paradigm used, the level of attention during the CS presentation has to be also taken into account (the excitatory and inhibitory trials were randomly scheduled in each session).

**ES/6**

SEX DIFFERENCES AND VOICE PRODUCTION IN STRESS CONDITION. *M. Nesic[1], M. Stankovic[2], V. Vuckovic[3], V. Nesic[4], S. Cekic[5]. [1,5] Institute of Physiology, Medical faculty, University of Nis, 18 000 Nis, Yu. [2,3] Electronic Engineering Faculty, University of Nis, 18000 Nis, Yu. [5] Faculty of Philosophy, University of Nis, 18000 Nis Yu*

The relationship of physiological correlates of stress and the changes in physical characteristics of speech signals is one of the main topics in the science of speech. Investigations on vocal indicators of stress indicate changes in parameters as fundamental frequency, intensity and duration. The aim of this experiment is to explore the changes of speech measures in the situation of exam. The sample was 100 students, one half were females and the other halves were males. They pronounced five vowels of Serbia language in relaxed condition and just used before exam. The pronunciations were recorded in the computer. Results showed changes in the fundamental frequency. The fundamental frequencies were between 156 Hz – 275 Hz for females and among 79 – 186 Hz for males. There were three groups of subjects: the first did not change fundamental frequency during experimental situation (N = 42), the second had lower fundamental frequency during experimental situation (N = 27), and the third had higher fundamental frequency (N = 31). The most females had lower or higher fundamental frequency in experimental situation (N = 37), but the most males did not change their fundamental frequency significantly (N = 29). Our investigation showed that, for the parameter fundamental frequency, there were individual differences in reaction to stress, and that there were sex differences, too.

## ES/7

**FEAR AND AGGRESSIVE RESPONSES IN NUCLEUS BASALIS MAGNOCELLULARIS-LESIONED RATS TREATED WITH VERAPAMIL.** *M. Popovic[1], K. Jovanova-Nesic[1], N. Popovic[1], D. Bokonjic[2], S. Dobric[2], and N. Ugresic[3]. [1] Immunology Research Center "Branislav Jankovic", Vojvode Stepe 458; [2] Medical Department, Military Technical Institute, Kataniceva 15 and [3] Department of Pharmacology, Faculty of Pharmacy, Vojvode Stepe 450, 11000 Belgrade, FR Yugoslavia*

It was found that excessive calcium influx represents the final common pathway of neuronal death. Therefore, the present study was done to investigate the effect of Ca-antagonist, verapamil (1.0, 2.5, 5.0 and 10.0 mg/kg sc, 30 min before the tests), on open field behavior and foot shock-induced aggression in male Wistar rats with bilateral electrolytic lesions of nucleus basalis magnocellularis (NBM). Bilateral electrolytic lesions of NBM induced: 1) significant increase of ambulation and number of inner squares entered and significant decrease of defecation in rats exposed to novel environment and 2) significant decrease of number of aggressive pairs, the number of fighting position and aggression scores as well as significant prolongation of the latency up to the first fighting position. Verapamil (2.5 and 5.0 mg/kg) significantly reduced ambulation and number of inner squares entered in NBM-lesioned rats. On the other hand, verapamil in dose of 2.5 mg/kg significantly increased defecation. In contrast to that, there is no restitution of aggressive behavior in NBM-lesioned rats after verapamil treatment. These results suggest that altered calcium homeostasis might play significant role in pathogenesis of experimental induced Alzheimer's disease. Administration of verapamil might successfully ameliorate disturbances of emotionality in novel environment, but without any influence on deficit of aggressive behavior appeared after lesions of NBM.

## ES/8

**HISTOCHEMICAL EXAMINATION OF CALCITONIN INFLUENCE ON THE HYPOTHALAMIC ACETYLCHOLINESTERASE ACTIVITY.** *P.Peric[1], S.Cekic[2]. [1,2]Dept.of Physiology, Nis University Faculty of Medicine, 18000 Nis, Yugoslavia*

Recent investigations have suggested that calcitonin(CT), a polypeptide hormone synthesized by thyroid gland's C-cell, has a neurotransmitting function in the central nerve system(CNS). The presence of specific and saturable CT-receptors were demonstrated within various brain structures with obvious enrichment in the hypothalamus. Some authors have indicated that in CT-treated rats brain 5-hydroxytryptamin(5-HT) synthesis and acetylcholinesterase(AChE) activity increased simultaneously. In our experiment, we try to explore possible CT-influence on the AChE-activity of hypothalamic nuclei. Experimental animals(male, adult guinea-pigs, weight between 300–400g) have been treated with CT (8 MRCunits/100g.b.w., i.m.) a half-hour before decapitation. At the same time, control group was treated with isotonic salt solution (1ml, i.m.) which was the solvent for CT. AChE activity has been identified and visualized histochemically by Karnovsky-Roots method. Comparing with the control group, the results showed various increase of the AChE activity in the hypothalamic nuclei, especially paraventricular nucleus, after the CT administration. Our experimental results are in agreement with postulated neurotransmitting function of CT in CNS, suggesting its involvement in the mechanism of cholinergic neuronal transmission. Whether or not the changes in the activities of the enzymes of the cholinergic system are directly mediated by CT or are secondary connected to the changes in other neurotransmitter systems such as 5-HT, remains to be determinated.

## ES/9

**EFFECT OF PHYSOSTIGMINE AND VERAPAMIL ON DEPRESSIVE BEHAVIOR IN AN EXPERIMENTAL MODEL OF ALZHEIMER'S DISEASE IN RATS.** *M. Popovic[1], D. Bokonjic[2], K. Jovanova-Nesic[1], N. Popovic[1] and S. Dobric[2]. [1] Immunology Research Center "Branislav Jankovic", Vojvode Stepe 458 and [2] Medical Department, Military Technical Institute, Kataniceva 15, 11000 Belgrade, FR Yugoslavia*

Our recent investigation indicate that the augmentation of central cholinergic activity with physostigmine and regulation of altered intracellular calcium homeostasis by verapamil could ameliorate cognitive disturbances in experimental model of Alzheimer's disease (AD). By using the model of learning helplessness, it was found that bilateral electrolytic lesion of nucleus basalis magnocellularis (an animal model of AD) in rats induced low level of depressiveness. The aim of this study was to investigate the effect of physostigmine and verapamil on learned helplessness reaction in rats after NBM lesions. Ten days after bilateral NBM lesions the physostigmine (0.045, 0.060 and 0.075 mg/kg s.c.) and verapamil (1.0, 2.5, 5.0 and 10.0 mg/kg s.c.) were administered 30 min before the test. The results indicate that bilateral lesions of NBM induced significant reduction of latency period up to first escape response. However, physostigmine in dose of 0.075 mg/kg and verapamil in doses 2.5 and 5.0 mg/kg significantly prolongated the escape latency period and thus produce a consolidation of depressiveness in NBM-lesioned rats. It could be concluded that both drugs exert a significant influence on depressive reaction in an animal model of AD.

## ES/10

**HIGH AND LOW RESPONDERS TO NOVELTY: DIFFERENCES IN MESOLIMBIC NORADRENALINE-DOPAMINE INTERACTION.** *T. Saigusa*, T. Tuinstra, N. Koshikawa* and A.R. Cools. Psychoneuropharmacology, University of Nijmegen, P.O. 9101, 6500 HB Nijmegen, The Netherlands. * Nihon University, Tokyo, Japan.*

Outbred strains of Wistar rats contain two types of rat: high responders (HR/APO-SUS) and low responders to novelty (LR/APO-UNS-US). Available pharmaco-behavioural studies suggest that the aminergic state of the n. accumbens differs between both types. The present study provides direct evidence in favour of this hypothesis. Microdialysis was used to assess (a) baseline concentration of dopamine (DA) in the n. accumbens of HR and LR sitting in their home cage (4 h after probe insertion), and (b) changes in this concentration during exposure to a novel cage. HR had a slightly higher baseline DA concentration than LR. In HR, novelty resulted in a large DA increase that was suppressed by the synthesis inhibitor a-methyl-p-tyrosine (MT: 0.1 mM, 40 min). In LR, novelty produced a small DA increase that was enhanced by MT. Next, we investigated the behavioural response to novelty after accumbens administration (0.5 ml/side) of reserpine (3 mg, given 24 h earlier) or MT (5 mg, given 5 min earlier) with or without the noradrenaline re-uptake inhibitor desipramine (DMI: 10 mg). MT changed the behaviour of LR into that of solvent-treated HR; this effect was counteracted by DMI. MT-treated HR showed behaviour that tended to go into the direction of that of solvent-treated LR; DMI potentiated this effect with the result that treated HR showed the behaviour of solvent-treated LR. These data suggest that the novelty-induced DA increase in LR/APO-UNSUS is due to the release of DA from reserpine-sensitive pools, being under inhibitory control of noradrenaline, and that the novelty-induced DA increase in HR/APO-SUS is due to the release of newly synthetized DA, being under stimulatory control of noradrenaline.

S 30

## ES/11
A RAT GENETIC MODEL INVESTIGATION OF THE NEURAL BASIS OF SENSATION SEEKING BEHAVIOR. *J. Siegel. Department of Psychology, University of Delaware, Newark, DE 19716, USA*

Sensation seeking and risk taking is a trait expressed in many behaviors, e.g. high-stakes gambling, dangerous sports, criminal acts and drug use. Individuals vary from being extreme risk takers to being excessively cautious. Using appropriate tests and observations, animals too can be evaluated for degrees of sensation seeking. Human and cat high sensation seekers show increasing amplitudes of visual evoked potentials (VEPs) to increasing intensities of light flash (VEP augmenting); low sensation seekers tend to show VEP reducing as a function of increasing flash intensity. P. Driscoll at Zurich has selectively bred rats for high vs. low active avoidance behavior, Roman High Avoidance (RHA/Verh) and Roman Low Avoidance (RLA/Verh) rats, respectively. The RHA/Verh rats exhibit behaviors comparable to those seen in human and cat high sensation seekers; the RLA/Verh rats exhibit low sensation seeking behaviors. Moreover, these two lines of rats show VEP augmenting and reducing, comparable to that seen in human and cat high and low sensation seekers; i.e., the RHA/Verh rats are VEP augmenters, the RLA/Verh rats are VEP reducers. These findings indicate we have a rat genetic model of an important behavioral trait (sensation seeking) that is accompanied by a biological- electrophysiological marker (VEP augmenting-reducing). We have demonstrated that the VEP augmenting or reducing recorded from cortex is not a reflection of augmenting or reducing at the lateral geniculate body; we showed that augmenting/reducing does not occur at the lateral geniculate. We conclude that augmenting and reducing are cortical phenomena. We now are investigating the role of neurotransmitters at the cortex that could modulate arriving visual messages up or down (augment or reduce) in high and low sensation seeking rats. Our first data show the role of glutamate receptor types responsible for the various components of the cortical VEP. Our particular interest is the component that augments or reduces (the early P1-N1 deflection). This early VEP component is generated by a non-NMDA excitatory amino acid receptor.

This work was supported in part by ARO Grant DAAL 03-88-K-0043.

## ES/12
THE EFFECTS OF ANGIOTENSIN II RELATED DRUGS ON THE APOMORPHINE STEREOTYPY IN RATS. *J. Tchekalarova and V. Georgiev. Lab."Experimental Psychopharmacology", Institute of Physiology, Bulgarian Academy of Sciences, 1113 Sofia, Bulgaria*

The present study was undertaken to evaluate the effects of the angiotensin II (ATII) receptor antagonists sarmesin (analogue of [Sar$^1$] ATII), where the tyrosine hydroxyl group in position 4 is methylated, sarilesin ([Sar$^1$, Ile$^8$] ATII) and Losartan (DuP 753), administered intracerebroventricularly (i.c.v.), on apomorphine (3 mg/kg i.p.) stereotypy in rats. Angiotensin II at doses of 1μg and 2μg (i.c.v.) significantly decreased apomorphine stereotypy but at a dose of 5 μg potentiated it. Sarmesin exerted dose-depentent biphasic effect on stereotypy, i.e. an increase at doses of 0.5 μg, 2 μg and 5 μg and a decrease at a dose of 10 μg. Sarmesin (5 and 10 μg), sarilesin (5 μg) and Losartan (100 μg) reversed the decreasing effect of ATII (2 μg) on apomorphine stereotypy. Thus, methylation of the hydroxyl group of Tyr$^4$ of [Sar$^1$] ATII produced an active analogue (sarmesin), which behaved as an ATII-receptor antagonist, whose effect on apomorphine stereotypy was more pronounced than that of sarilesin, a peptide ATII receptor antagonist. In conclusion, all the three drugs tested (sarmesin, sarilesin and Losartan) influenced apomorphine stereotypy and reversed the effect of ATII on stereotypy. The results also suggest evidence for ATII-DA interactions in brain structures related to apomorphine stereotypy.

*Acknowledgements.* This study was supported by the European Community, through the Copernicus programme, contract No CIPA CT 94-0239 and by Grant L-526 from the National Fund "Scientific Research" at the Bulgarian Ministry of Education, Science and Technology.

## ES/13
MATERNAL DEPRIVATION IN THE BROWN NORWAY RAT RESULTS IN LONG-TERM ALTERATIONS OF THE CORTICOSTEROID SYSTEM AND BEHAVIOR. *J.O. Workel, A. Ledenboer, E.R. de Kloet and M.S. Oitzl. Leiden/Amsterdam Center for Drug Research, Division of Medical Pharmacology, University of Leiden, P.O.Box 9503, 2300 RA Leiden, The Netherlands*

Early life experience by maternal separation may exert life long alterations in the neuroendocrine and behavioral stress response. The main regulators of the stress response are corticosteroid hormones secreted by the adrenal gland. Corticosterone (CORT), the main corticosteroid in rats, exerts its action via binding to high (MR) and low affinity receptors (GR) in the brain, controlling its own release by feedback action and thereby maintaining homeostasis and adaptation to stress. During a certain period in the first weeks after birth, corticosteroid levels are low in the rat, even after stress. Maternal presence has been shown to be the inhibiting factor for this phenomenon. This period, the stress hyporesponsive period lasts from postnatal day 4–14, and is very important for normal brain development. Therefore, neonatal manipulations interfering with the corticosteroid system, may modulate brain functioning in adulthood and may have consequences for the aging process.

Our project is designed to study the effect of maternal deprivation (single 24 hour mother-pup separation at postnatal day – pnd – 3.) and neonatal handling (15 min daily separation mother-pup from pnd 7 to 11) in later life: at pnd 18, and 3, 12, 24 and 36 months of age. We use the Brown Norway rat, known for its longevity and healthy lifespan, and analyze basal and stress-induced ACTH and CORT in blood, MR and GR mRNA in the hippocampus, and GR and CRF mRNA in the hypothalamus, and behavior in the Morris water maze. Corticosteroids are known to influence spatial learning and memory. – Maternal deprivation at pnd 3 for 24 hours results (i) at pnd 18 in low basal, but enhanced stress-induced ACTH release. CORT, CRF, MR and GR mRNA are not influenced; (ii) at 3 months in reduced stress-induced CORT release, while ACTH and CRF mRNA are not affected. These animals are impaired in the acquisition of the spatial task. (iii) at 12 months in an enhanced ACTH release, reduced CRF mRNA levels, increased MR mRNA in the dentate gyrus of the hippocampus. In neonatally handled animals, MR mRNA is increased in all hippocampal subfields. GR mRNA levels are not influenced by any maternal separation procedure. In the water maze, maternally deprived animals take longer to acquire the task. Neonatally handled animals take more time to locate the platform in the first trial after a free swim trial. Non-deprived animals displayed a more efficient search strategy.

Taken together, the elements of the corticosteroid systems are differentially affected at the three ages we looked until now. At pnd 18, the pituitary appears to be more reactive to stress. At 3 months the adrenal cortex is hyporesponsive to novelty stress and at 12 months the MR mRNA expression is increased in deprived and handled animals. Adult animals which have been maternally deprived are impaired in the acquisition of the spatial task. At present we can state that the frequency and duration of maternal separation during infancy differentially affects the stress-responsiveness and learning for at least one year.

Supported by the Dutch Organization for Scientific Research NWO grant 554-545.

**ES/14**

THE HYPOTHALAMO-NEUROHYPOPHYSEAL SYSTEM – A MODEL FOR STUDYING BASIC PRINCIPLES OF NEUROPEPTIDES AND THEIR BEHAVIORAL IMPACT: *C.T. Wotjak, R. Landgraf, M. Engelmann: Max-Planck-Institut für Psychiatrie, Kraepelinstr. 2, D-80804 München, BR Deutschland*

Although more than 70 neuropeptides had been identified in the mammalian central nervous system, surprisingly little is known about the basic principles, dynamics and behavioral impact of their central release. This is mainly due to the low concentration of the neuropeptides in the brain tissue which makes it difficult to measure their release within distinct brain areas concomitantly with behavioral testing. Thus, typically plasma peptide levels are used as indicators for the secretory activity of the central components of the respective neuropeptidergic system. In this context, the two nonapeptides vasopressin (AVP) and oxytocin (OXT) of the hypothalamoneurohypophyseal system provide an interesting model for comparing central and peripheral release patterns as they are synthesized in relatively high amounts within the hypothalamic supraoptic (SON) and paraventricular (PVN) nuclei and released into their extracellular fluid. These peptides can thus be measured not only in blood samples but also in dialysates/perfuates collected from the SON and PVN. Using both the microdialysis technique and chronically implanted jugular venous catheters, we were able to demonstrate that a 10-min forced swimming session is potent to trigger the release of both AVP and OXT within the PVN and SON of adult male rats. The central release of OXT was paralleled by the secretion of this peptide into blood. Interestingly, AVP levels in plasma remained unchanged during swimming. The latter finding indicates that vasopressinergic neurons are capable of releasing the peptide in a dissociated manner from somata/dendrites (within SON and PVN) and axon terminals (from neurohypophysis into blood). These data suggest the hypothalamo-neurohypophyseal system not only as a model for studying common and different principles underlying somatic/dendritic and axonal release but also intrahypothalamically released AVP and OXT to be causally involved in the behavioral response of the organism to emotionally and physically challenging situations. Furthermore, our findings provide evidence that plasma hormone levels do not necessarily reflect the secretory activity of central components of neuropeptidergic systems and, thus, attempts in correlating plasma peptide levels with behavioral performance need to be interpreted with caution.

# Functional Neuroanatomy and Functional Neurophysiology

## FN/1

CYTOCHROME OXIDASE STAINING IN HUMAN PRECU-NEUS, SUPERIOR PARIETAL LOBULE AND INTRAPARIE-TAL SULCUS: FURTHER ARGUMENTS FOR FUNCTION-ALLY DIFFERENT AREAS. *Anne-Claude Cottier Eskenasy, Gabrielle Di Virgilio and Stephanie Clarke. Institut de Physiologie, Université de Lausanne, Rue du Bugnon 7, 1005 Lausanne, Switzerland*

The superior part of the posterior parietal lobe is known to receive multimodal input. Lesion studies indicate that it is involved in a wide range of functions such as visuomotor coordination, visual and tactile recognition, visuo-constructive and visuo-spatial abilities and praxias. Recent activation studies have shown that discrete parts of the parietal cortex can be selectively activated by specific tasks. According to von Economo and Koskinas (1925) this whole region consists of a single architectonic area PE with local differences ( PEm, PEp, PE(D), PEgamma ); Brodmann (1909) distinguishes 2 areas (5 and 7).
We investigated this region on post-mortem human brain tissue using cytochrome oxidase (COX), acetylcholinesterase (AChE) and NADPH-diaphorase (NADPHd) staining. Indeed, in previous studies on the human cortex, COX was found to be a useful marker for extrastriate visual and auditory areas and AChE for auditory versus speech areas.
The pattern of COX, AChE and NADPHd staining was studied in the precuneus, superior parietal lobule and intraparietal sulcus in 6 normal hemispheres. Aged respectively 70, 71 and 90, the subjects died from acute cardiac problems and had no known neurological diseases. Time between death and fixation varied between 8 and 12 hours. The brains were fixed by perfusion with 4% paraformalde-hyde through the 3 main arteries. The parietal, temporal and occipital lobes were cut frozen in serial coronal sections, stained in alternation for COX, NADPHd, AChE, Nissl and myelin.
COX was revealed in individual cortical neurons and in the neuropil. The intensity of staining varied radially and tangentially within the cortex. The laminar distribution of COX was characterized by either one dark band in layers III-IV or two in layers III and V respectively. Two cortical regions were prominent: i) a darkly stained one-band region in the intraparietal sulcus; and ii) a region with a two-band pattern on the precuneus and the medial part of the superior parietal lobule. These regions measured ca $1cm^2$ and $2-3cm^2$ respectively. Slight variations in the myelin and AChE patterns coincided with the different COX regions.
The present results argue for the cortex of the precuneus, the superior parietal lobule and within the intraparietal sulcus to contain several, most likely functionally distinct areas, as suggested by clinical data.

## FN/2

ALTERATION IN THE EXPRESSION OF NOS-IMMUNO-REACTIVITY AND NOSmRNA IN THE SPINAL CORD IN RESPONCE TO VARIOUS MODES OF ELECTRICAL STIMULATION OF THE SCIATIC NERVE IN THE RAT: *A. Kaske, U. Hoheisel, L. Klimaschewski, A. Reinert, S. Mense. Institut für Anatomie und Zellbiolgie, Im Neuenheimerfed 307, D-69120 Heidelberg*

Sensitization in responce to peripheral noxious stimuli and windup in responce to electrical stimulation of spinal afferents are models of spinal neural plasticity. Both models involve the activation of NMDA-receptors and the enzyme nitric oxide (NOS), which has been shown to play a crucial role in the process of LTP in the hippocampus. The structure and regulation of NOS is rather complex and involves 5 cofactors. Most important for its activation is the dimerisation of the enzyme and contact to Ca-calmodulin in a Ca++ dependent way. Increase of intracellular Ca++ by the activation of NMDA-receptors is followed by NOS-activation. NO triggers further second messenger cacades via activation of the soluble guanylate cyclase.

The conditions in the spinal cord are unique that only unmyelated small calibre afferents – C-fibres – release neuropeptides like SP which is known to deblockade NMDA-receptors. Myelinated A-fibres release probably glutamat. Our model allows to stimulate all fibres including C-fibres and myelinated fibres separately. The sciatic nerve was stimulated while the compound action potential was monitored from the suralis nerve to control selective stimulation. NOS-immunoreactivity and NOSmRNA expression (only evaluated in L3 after 2h) of the lumbar segments L3 as a marginal target of sciatic afferents and the segement L5 as the main target were evaluated.
At 1 Hz stimulating frequency (within the windup regime) stimulation of A-fibres led to a monophasic increase of the number of NOS-immunoreactiv e neurons in the superficial dorsal horn after 2 h, but was not accompanied by changes of NOSmRNA expression. Whereas stimulation at C-fibre intensity was followed by a biphasic response – an inititial increase after 1–2 h followed by a decrease below control level (transitory in L5) after 4 h – and showed altered expression of NOSmRNA after 2h. The data point to a restricted availibility of NO for neuroplastic changes under maximal C-fibre stimulation after 2–4 h. This coincides with data gained from hippocampal slices which point to a general time window of 2–3 h in which cascades of enzyme activation and protein synthesis (involving gene activation) have to take place to maintain induced LTP.In the context of LTP NO has been discussed as a retrograde messenger, pointing to a general interplay of NO, NOS enzyme induction and LTP within a restricted time window. This suggests that a cell has to go through a well defined sequence of events (program) involving second messenger cascades and gene-induction to lay down permanent traces in the neural network.

## FN/3

SIMULATION OF THE SELF-CONTROL OF CORTICAL ACTIVATION BY COORDINATION OF TOP-DOWN AND BOTTOM-UP SELECTION PROCESSES INVOLVING SUBCORTICAL LOOPS AND THE HIPPOCAMPUS FORMATION: *A. Kaske, Institut für Anatomie und Zellbiologie III, Im Neuenheimer Feld 307, D-69120 Heidelberg*

The projections between cortical areas can be classified according to their microscopic termination patterns into the classes of feedforward (FF)- and feedback (FB) projections, implicating heirarchical relations between interconnected cortical areas. Additionally areas with strong reciprocal connections project to the same subcortical targets – i.e. the striatum – in an overlapping fashion. The order of cortical FF- and FB-patterns is reciprocated in the fine structure of the projections to the striatum. Top-down projecting FB-areas terminate in the dopamin (DA)-rich striosom compartement of the striatum, while bottom-up projecting FF-areas terminate in the GABA-rich compartement of the matrix. Efferents (GABA) of the striosom form a loop with DA-containing efferents from the substantia nigra, while the efferents of the matrix form a loop: matrix-globus pallidus-thalamus FF-cortex-matrix. The later loop is modulated by the first loop, which is controled by the FB-area (higher order area). The first simulation designed to represent this interaction takes account of the known anatomy – i.e. the striosom/matrix dichotomy of the striatum with the complementary distribution and function of the D1 and D2 dopamin receptors in the striatum. The different time constants of EPSPs, IPSPs and second messenger effects (DA) were included in the model. The simulation demonstrates that the FB-area actually controls the loop of the FF-area by setting its gain and ground level of activity. The interaction has spatiotemporal bandpass properties spatiotemporal correlations in that bandwith. Deactivation of the substantia nigra in the model reproduces the symptoms of M. Parkinson (loss of gain/ akinesia; decalibrated ground level/ rigor). The implication of the model replanted into the actual anatomy is that a higher order FB-area controls the activty of FF-areas (5–10) to which it is connected. This mechanism amounts to a top-down control of cortical activtiy.
The hippocampus formation (HC) is reciprokely connected to the cortex, FB-areas (prefrontal) tend to project to a slab of the dentate gyrus (DG), while the FF-areas (parietal and temporal) connected to

this FB-area tend to project to a slab of the entorhinal cortex (EC). Corresponding DG and EC are interconnected by a loop involving a slab of the hippocampus proper (CA1-4 and subiculum). The HC was simulated on 2 scales of organisation, as macroscopic map and on a complementary micro-scale. For the map the known anatomy was incorperated into the model emphasising the anisotropy of the fiber tracts and the relative extension of excitatoric axon arbors (granule cells and pyramids) and inhibitory axon arbors (basket cells). The micro-network simulation focused on processes below the scale of the inhibitory axon arbor.The simulations (on both scales) demonstrate that the HC acts as a comparator of the (spatiotemperal) activity-pattern of interconnected FF- and FB-areas. It signales match by stabilising the ongoing activity with its own spatiotemporally defined activity. A mismatch results in chaotic activty of the HC. This chaos spoiles the pattern of activity in the cortex. The spread of activity is filtered in the cortex in that sense that only correlated activity is amplified by the convolution of the striatal loops. The new pattern of cortical activation is again cross-checked by the HC. This leads to cycle of top-down selection of activity followed by a bottom-up propagation of the HC-dynamics until a stable match is reached.

## FN/4

DIFFERENT STAGES OF CORTICAL PROCESSING REVEALED BY SUBCOMPONENTS OF SENSORY EVOKED POTENTIALS. *E. Kublik, P. Musia P. and A. Wróbel. Department of Neurophysiology, Nencki Institute of Experimental Biology, 3 Pasteur Str., 02-093 Warsaw, Poland*

Five unanesthetized rats with chronic electrodes implanted in the barrel cortex were used in this study. In the beginning of the experiment animals were accustomed to rest in a plexiglass tube with head restrained in a holder. After implantation of electrodes the chosen vibrissa was stimulated with piezoelectric device and evoked potentials (EP) were recorded from the barrel cortex. Five habituation sessions were followed by a conditioning session in all of which the animal received 100 vibrissa stimulations with intervals randomly scattered from 30 to 45 s. In the conditioning session the first 30 vibrissal stimulations allowed for stabilization of the EPs. All remaining stimulations were followed by a mild electric shock (unconditioning stimulus) applied with a 250 ms delay to the ear on the same side. The whole conditioning session lasted for about an hour.

The first negative component (N1) of EP consisted of two peaks which differed by 1.5–2 ms in latency. The contribution of these subcomponents of N1 to the integral value of EP was calculated within the 5 ms period containing both peaks. This procedure allowed to classify EPs with respect to relative amplitude of the two subcomponents. The second class differed from the first one by increased amplitude of the second subcomponent. Introduction of the conditioning procedure changed the control ratio of the two classes in such a way that number of EPs with the enhanced amplitude of later subcomponent rapidly increased.

We hypothesize that the two N1 subcomponents might reflect the successive stages of sensory information flow within the barrel cortex. The conditioning procedure would recruit larger population of cells at the higher processing level (delayed by one synapse) and thus enhance the amplitude of the second subcomponent.

## FN/5

ON THE ETHNIC PECULIARITIES OF CEREBRAL LATERALIZATION. *Malkhas Makashvili, Ia Samadashvili, Shayantan Shayi. Institute of Physiology, Tbilisi, Georgia*

McManus and Humphrey (1973) explored the fact that the European portrait painters of XIX–XX centuries have tended to paint human profiles, oriented towards the left rather than towards the right. According to Hufschmidt (1980) the preference for the left profile direction is traced back to the early Greek period in paintings, coin portraits, cameos and vase portraits. Before this time the Assyrian, Egyptian and Sumerian cultures faced more profiles to the right.

The profile shift from right to left occurs in the early Greek period and is related to a shift in script and in letter profile at the same time. According to Hufschmidt (1980, 1988) it was during this time that the association of speech dominance with the left hemisphere and the control of movements by the right became established.

The hypothesis gives rise to special questions: If a shift in painting and script direction has been caused by the development of cerebral lateralization, what are then the peculiarities of profile painting and brain functional asymmetry in oriental human populations, which still write from right to left?

Do oriental and European ethnic groups differ in functional capacities of the right and the left brain hemispheres?

Georgian students at the Tbilisi Medical University were requested to draw a human profile, choosing the profile direction on their own. The same instruction was given to 36 guest-students from South Asia (Srilanka, India, Pakistan, Nepal).

Of 100 Georgian students 93% directed profile towards the left and the other 17%-towards the right. No difference was find between the groups of male and female subjects.

Thus, 76% prevalence of face directions towards the left occurs in the group of Georgian subjects.

On the other hand 50% of guest-students directed profile towards the left, the other 50%-towards the right. Thus, subjects from South Asia do not show the preference for one of two profile directions. Two subjects in the "leftward" group and two students in the "rightward" group have their native script direction from right to left. Other subjects, as well as Georgian students, have their native script direction from left to right.

Hence, the script direction does not influence the direction of profile drawing.

Evidence are convincing (Kinsbourne, 1972, Nikolaenko, 1993) that tendency to draw profiles oriented towards the left is due to the specific involvement of the right cerebral hemisphere in the process of drawing, whereas the rightward profile direction is attributable to the predominant activity of the left hemisphere.

The results of the present study suggest that ethnic groups differ in hemispheric specialization for profile drawing.

## FN/6

MACROGRAPHIA AND LATERALITY. *M. Rümbeli, C. Roehrenbach, P. Brugger, M. Regard. University Hospital, Neuropsychology Unit, CH-8091 Zürich, Switzerland*

Aim: After we observed an association between elevated mood and oversized copies of figures (macrographia) in patients with lesions predominantly involving the right anterior cortical regions, we investigated hemispheric differences in size perception and graphomotor expression in healthy subjects.

Methods: 40 right-handed subjects (20 f, 20 m) participated in two lateralized tachistoscopic studies and a drawing task. Stimuli were simple grey squares differing in vertical or horizontal extension. Two squares, which were identical or differing with respect to size, were simultaneously presented one to the left (LVF) and one to the right visual field (RVF) in a random order for 80 ms. Forced-choice correct responses, „pseudohits" and response latencies were analyzed. Additionally, each subject copied the Rey complex figure twice, 50% first with the right hand and 50% first with the left.

Results: Independent of orientation, size differences were significantly faster and more accurate in the LVF. The same result was found when stimulus pairs of equal size were presented. The left hand copies were bigger than the right-handed drawings, but, statistically, the differences in height, width and area were not significant. There were no gender differences.

Conclusions: The findings of a LVF superiority for judging stimuli of different and same size and of bigger drawings with the left hand suggest that the right hemisphere not only perceives size differences of simple stimuli more correctly than the left hemisphere, but also falls for size illusions. We conjecture that the macrographia in patients (who draw with their dominant right hand) is a compensatory effect of a lesioned right hemisphere.

**FN/7**

FUNCTIONAL NEUROANATOMY OF ALERTNESS. A PET-STUDY: *Walter Sturm, Klaus Willmes, Anna DeSimone, Volker Hesselmann, Carsten Specht, Hans Herzog, Bernd Krause (University Clinic Aachen, Research center Jülich)*

In 14 normal subjects a study of the functional neuroanatomy of alertness was carried out by means of a $^{15}$O-PET-activation. The paradigm for the alertness activation was a simple visual reaction time task. The subjects had to respond as fast as possible by pressing a response key with their right thumb whenever a white central light spot appeared. Control conditions were a rest condition with passive fixation of a central fixation point and a combined sensori-motor control condition, during which the response key had to be pressed arbitrarily while watching a fickering central light spot. The difference between alertness and combined control conditions revealed a significant right hemisphere activation in the anterior cingulate gyrus, in the white matter adjacent to the dorsolateral prefrontal cortex, and in the reticular formation at the dorsal ponto-mesencephalic transition area (possibly at the location of the locus coeruleus). The results corroborate the special role of a right frontal network in the control of brain stem induced activation of attention (alertness).

**FN/8**

PROCAINE INACTIVATION OF THE PEDUNCULOPONTINE TEGMENTAL NUCLEUS SUPPRESSES HIPPOCAMPAL THETA RHYTHM IN RATS. *W. Trojniar, A. Nowacka, J. Tokarski. Department of Animal Physiology, University of Gdańsk, 24 Kładki St., 80-822 Gdańsk, Poland*

It is well established that the hippocampal theta rhythm is driven by activation of several sites in the lower brainstem diencephalon and prosencephalon. Out of the brainstem structures, the nucleus pontis oralis of the rostral pontie reticular formation (RPO) was proved to be directly involved in the generation of the theta. Brainstem microinjections of a cholinergic agent, carbachol, besides RPO evoked robust theta activity also from the area of the pedunculopontine tegmental nucleus (PPN) (Vertes et al., 1993). This raises a question of the relative importance of PPN in the generation of the theta.

The objective of the present experiment was to assess the effect of temporal inactivation of PPN on the hippocampal theta rhythm evoked in anaesthetized rats by sensory stimulation.

The experiment was done on urethane anaesthetized male Wistar rats implanted with hippocampal recording electrodes in the stratum moleculare of the dorsal blade of the dentate gyrus and with an injection cannula unilaterally in the region of PPN. Theta rhythm of the amplitude of at least 500 mV was evoked by sensory stimulation (tail-pinch) several times in the preinjection conditions and in 10 min intervals after intra-PPN injection of 1 ml of 20% procaine.

It was found that sensory-elicited theta completely disappeared for up to 30 min after procaine injection. It recovered parallelly to the drug wash-out. The effect was anatomically specific, not present after control procaine administration to the neighbouring tissue.

The results indicate that PPN may play a key role in the brainstem control of hippocampal theta activity.

# Learning and Memory

## LM/1

MALE FEMALE DIFFERENCES IN RATS' SPATIAL STRATE-
GIES AND CHOLINERGIC FUNCTION: *C. Brandner\*, F. Schenk
- Institut de Physiologie, CH-1005 Lausanne*

The majority of the studies concerning sex differences report that
males perform better than female in a variety of spatial tasks. The
cholinergic system also appears to show sexual dimorphism in the
rat hippocampal formation. Thus, it is unclear whether sex differences in spatial tasks are only secondary to the development of
qualitatively different learning strategies or whether they are due
to functional differences of specific neural mechanisms involved
in the elaboration of spatial representation. On the one hand, sex differences in exploratory patterns or in the encoding of visuospatial
information might modify the development and maintenance of a
spatial representation. Males are reported to attend to geometric cues
while female attend to both geometric and landmarks cues during
spatial tasks. On the other, female spatial performance appears sometimes more sensitive to anticholinergic drugs administration,
and sometimes more resistant to cholinergic lesions or pharmacological manipulations. This suggests functional differences in the
cholinergic system.
We have found in different experiments that spatial memory was
particularly impaired in subjects with reduced cholinergic activity
when they had been trained in the presence of a salient local cue.
To account for these effects, we propose that an allocentric coding
of spatial relations requires a process of compensation for differences in the salience of spatial cues, that might be critically dependant on the activity of the cholinergic system. The question then
is to what extent a difference in the activity of the cholinergic system might account for sex differences in cued spatial tasks.
We analysed the effects of cholinergic manipulations on spatial performance by Long Evans and PVG female rats in a Morris Navigation Task. Perinatal choline administration prevented the overshadowing effect produced by the presence of a salient local cue in adult
rats. Postnatal exposition to icv injections of NGF decreased drama-
tically the number of errors made by adult female rats in a radial arm
maze. Finally, the overshadowing effect produced by the visible escape platform was age and sex dependant. We propose that the sexual dimorphism observed in these spatial tasks is due to the use of
different spatial strategies associated with different spatial representations. These strategies seem to be related on a difference in the coding of spatial information, females making a more flexible use of
multiple cues than males.

## LM/2

THE NEUROPSYCHOLOGY OF "PARANORMAL" BELIEF. *P.
Brugger[1], M. Regard[1], T. Landis[2], R.E. Graves[3]. [1] Neurology Clinic,
University Hospital Zurich, CH-8091 Zurich, Switzerland. [2] Neuro-
logy Clinic, University Hospital Geneva, CH-1211 Geneva 14, Swit-
zerland. [3] Dept. of Psychology, University of Victoria, Box 3050,
Victoria, British Columbia, Canada V8W 3P5*

Belief in a "paranormal" causation of coincidences is widespread
and largely independent of psychosocial, including educational, factors. We present two experiments that suggest a neuropsychological
basis for belief in the paranormal, disinhibition of semantic-associa-
tive processing. In Exp. 1, students were asked to guess a random
series of events, some of them being associatively related to one another. Believers in extrasensory perceptions more strongly avoided
consecutive guessing of semantically related response alternatives
than did nonbelievers. In Exp. 2, students were asked to spontaneously generate as many words as possible that would belong to either one of two semantically distinct categories (animals, and fruit).
Believers in paranormal forms of causation, overall generating as
many words as nonbelievers, were found to shift more frequently between the categories. These results show an association between belief in the paranormal and (1) a relative overrepresentation of semantic similarity, and (2) a relative preference for remote (beyond-ca-

tegory) over close (within-category) associations. Our finding is relevant for a neuropsychological interpretation of delusional beliefs
which assumes a disinhibition of associative processing, presumably
due to temporal-limbic hyperactivity.

## LM/3

THE CONTRIBUTIONS OF MEDIAL DIENCEPHALIC AND
FRONTAL LOBE REGIONS TO MEMORY. *P. Calabrese[1], H.J.
Markowitsch[2], D.Y. von Cramon[3] and A.G. Harders[4]. [1] Clinic of
Neurology (Knappschaftskrankenhaus), Faculty of Medicine, Uni-
versity of Bochum, D-44892 Bochum, Germany. [2] Physiological
Psychology, University of Bielefeld, D-33501 Bielefeld, Germany.
[3] Max-Planck-Institute of Cognitive Neuroscience, D-04103 Leipzig,
Germany. [4] Clinic of Neurosurgery (Knappschaftskrankenhaus),
Faculty of Medicine, University of Bochum, D-44892 Bochum, Ger-
many*

Sixty-one bi- or unilaterally brain damaged patients, divided into 3
diencephalic, 3 dorsolateral prefrontal and an orbitofrontal/basal forebrain damaged group were studied and compared with each other
using a major test battery which was comprised of tests measuring
attention, intelligence, verbal and nonverbal short and long-term memory, procedural memory and priming, interference, and remembr-
ance of emotional versus neutral material. Overall, all groups had
moderate to severe memory problems which were usually more severe for the bihemispherically damaged groups.
Between groups, the diencephalic patients showed the widest range
of severe memory deteriorations, followed by the orbitofrontals.
Left diencephalic damage had much more devastating consequences
than right hemispheric. The orbitofrontals manifested particular problems in binding cognitive and emotive components of memory, as
measured by comparing their performance in texts and photographic
scenes varying in emotional content. Of the dorsolateral prefrontals
only the bilaterally damaged had some working memory deteriorations, while all prefrontals performed quite well in tests of attention
and concentration, when compared to the other groups. Somewhat
unexpected results were that the prefrontals as a group did show
the release from proactive interference effect and that the diencephalics and orbitofrontals were impaired in priming performance.
Our data indicate that the memory problems of dorsolateral prefrontals are secondary to other cognitive deficits; memory deficits of the
orbitofrontals are more similar to those of the diencephalics, but in
general less severe and more closely related to the emotive dimension of information processing. It is argued that the basal forebrain
structures' contribution to memory processing is more of a suppor-
tive, facilitating nature, guiding the emotional evaluation of new stimuli and thereby influencing the likeliness of new binding processes
and consequently of the consolidation of new long-term memories.
Furthermore, portions of the orbitofrontal cortex seem to be engaged
in the effortful retrieval of information and in retrieval attempts,
while the medial diencephalic structures constitute essential gate-
ways or bottleneck structures for final memory storage.

## LM/4

EFFECT OF INTRAAMYGDALA ADMINISTRATION OF CO-
CAINE ON PASSIVE AVOIDANCE BEHAVIOUR IN CD1
MICE: INVOLVEMENT OF THE DOPAMINERGIC SYSTEM.
*V. Cestari[1,2], A. Ciamei[1], C. Castellano[1]. [1] Istituto di Psicobiologia
e Psicofarmacologia (CNR) – via Reno 1, 00198 Roma, Italy. [2] Dipartimento di Genetica e Biologia Molecolare, Università degli Stu-
di di Roma "La Sapienza" – P.le Aldo Moro 5, 00185 Roma, Italy*

In the present research we have investigated the effect of intraamygdala posttraining injections of cocaine on memory consolidation in
mice. We have also studied the role of the dopaminergic system
in the effect of cocaine. For this purpose, the animals were posttraining injected into the structure with a combination of cocaine and of
the D1 or D2 dopamine (DA) receptor agonists SKF 38393 (SKF)
and LY 171555 (LY). The mice were bilaterally implanted with gu-
ida cannulae on the skull, at the coordinates corresponding to the

central nucleus of the amygdala, by means of a stereotaxic Narishige apparatus. Seven days following surgery they were trained in a passive avoidance test and immediately after injected into the amygdala with an injection cannula connected with a Hamilton microsyringe. 2 ml of drug solution were administered for each side. The animals were tested 24 hours after injection. The drugs used were: cocaine (2.5 or 5 mg), SKF (1 or 2 mg) and LY (0.5 or 1 mg). The results showed a retention improvement following cocaine (5 but not 2.5 mg), SKF (2 but not 1 mg) and LY (1 but not 0.5 mg) administration, as compared with saline injected mice. Moreover, by themselves ineffective doses of SKF and LY enhanced the facilitation induced by 5 mg of cocaine.

Some experiments have shown retention enhancement in rats and mice, tested in active and passive avoidance tasks respectively, following posttraining intraperitoneal (i.p.) administration of cocaine, SKF and LY (Introini-Collison and McGaugh, 1989; Castellano et al., 1991; Janak et al., 1992). Moreover in mice tested in a passive avoidance situation, i.p. administration of the D1 or D2 DA antagonists SCH 23390 and (-)-sulpiride respectively, antagonized the memory improvement exerted by cocaine (Puglisi-Allegra et al., 1994). Finally it has been recently shown that the positive effect of cocaine on memory consolidation in mice tested in a passive avoidance task, is blocked by lesions to amygdala (Cestari et al., 1996).

The results of the present research confirm that amygdala is a critical structure in mice for the effect of cocaine in memory to become evident. They show moreover that SKF and LY improve memory by acting on the central nucleus of the amygdala, and that this region is also involved in the interaction between cocaine and the dopaminergic system in CD1 mice tested in a passive avoidance condition.

## LM/5
EVENT RELATED POTENTIALS DURING A CATEGORICAL LEARNING TASK. *M.F. El Bab, B.J. Colleypriest, E.M. Sedgwick. Clinical Neurological Sciences, University of Southampton, U.K.*

Event related potentials were recorded from 25 scalp electrodes on 15 normal subjects. Our aim was to determine whether there was a difference in potential in those who learned compared with those who did not. Computer generated patterns had to be classed as type A or B by pressing the mouse buttons. Each pattern presentation was randomly generated but obeyed the rule for its type, A or B. Clues from border effects or contrast change were eliminated.

In 200 trials, 7 subjects learned (> 70% correct) and 8 did not. The first 35 responses and last 35 were averaged and the mean amplitudes during two windows (WI 380-580 msec; W2 700–950 msec) were measured.

Learners showed an increase in positivity during W1 and W2, whereas non learners showed increased positivity during W2 only. The increased positivity was greater over the right hemisphere than the left.

It is concluded that successful learning is associated with an increase in late positivity. Also, that categorical learning is predominantly a right hemisphere activity.

## LM/6
CONTRARY TO ANTERIOR CINGULATE CORTEX, PRELIMBIC CORTEX IS INVOLVED IN WORKING MEMORY PROCESSES BUT NOT NECESSARILY IN SPATIAL WORKING MEMORY. *P. Gisquet-Verrier, B. Delatour. Laboratoire de Neurobiologie de l'Apprentissage et de la Mémoire, URA C.N.R.S. 1491, Université Paris-Sud, 91405 Orsay, FRANCE*

The prelimbic area (PL) of the medial prefrontal cortex (mPFC) in the rat has strong anatomical connections with the hippocampus and other related limbic structures, whereas the anterior cingulate cortex (ACd) is considered to be a premotor-type cortex. We have investigated the effects of specific lesions of the PL and of the ACd cortex on various behavioral tasks. From these results, it appears that the PL cortex is involved in delayed response tasks where the response depends on specific information delivered previously (i.e. working memory), while ACd cortex is involved in the acquisition of tasks requiring response patterning involving fixed sequential behavior, such as spatial delayed alternation.

As large mPFC lesions have often been involved in the acquisition of spatial information, it was of interest to investigate the respective role of PL and ACd cortices in the acquisition of a spatial working memory task, such as the radial maze task.

Rats with specific ibotenic acid lesions of the PL or the ACd cortex were compared with sham operated rats during the acquisition of a standard radial arm maze task and no difference was observed between the 3 experimental groups. During a second phase, each daily trial was divided into two phases. During the first phase, rats was allowed to freely visit four different arms among the 8 baited arms of the radial maze. The trial was then interrupted and the rat was placed in a waiting box for 1 min before being placed on the radial maze again until the four remaining baited arms were visited. Under these conditions, PL rats were transiently disrupted but rapidly recovered the level of performance shown by the two other groups. Extending the time delay from 1 to 5 min had no disruptive effect, however, extending it further from 5 to 30 min resulted in a transient disruption of performance in the PL rats.

In all, these results indicated that ACd cortex is not involved in the acquisition of complex strategies when sequential organization of the responses is flexible. PL cortex does not seem to be necessary for the working memory processes required to performa standard radial arm maze task correctly but it is involved, at least transiently, in another kind of working memory, necessary to retain which set of arms have been visited, over time intervals longer than those required for the ongoing behavior. These results suggest that the hippocampus might be sufficient to subserve spatial processing function and related working memory capacities, whereas the PL cortex would be more so concerned with some form of specialized working memory functions and particularly those involved in recency judgement.

## LM/7
ANXIOLYTIC ACTION OF NEUROKININ SUBSTANCE P ADMINISTERED SYSTEMICALLY OR INTO THE BASAL FOREBRAIN. *R.U. Hasenöhrl; O. Jentjens; M.A. De Souza Silva; C. Tomaz\*; J.P. Huston. Institute of Physiological Psychology, University of Düsseldorf, Universitätsstr. 1, D-40225 Düsseldorf, Germany; \*Department of Psychobiology, University of Sao Paulo, Ribeirao Preto, 14049 Brazil*

The neurokinin substance P (SP) plays a role in reinforcement and memory. Reinforcing and memory-promoting effects of SP were found upon injection into several parts of the brain and in the periphery. Given the close link between fear/anxiety and memory processes for negative reinforcement learning, the aim of the present study was to gauge the effect of SP in the rat elevated plus-maze (EPM) and social interaction test (SIT) of anxiety. SP was tested at injection sites where the neurokinin had been shown to promote learning and to serve as a reinforcer, namely in the periphery (after IP injection) and in the region of the nucleus basalis magnocellularis (NBM). In the first experiment, SP was injected IP in 4 doses ranging between 5 and 500 mg/kg immediately before the rats were tested on the EPM for 5 min. The treatment with SP at 50 mg/kg elevated the time spent on the open arms and excursions into the end of the open arms and increased scanning over the edge of an open arm, indicative of an anxiolytic-like action; at the high dosage of 500 mg/kg, SP decreased the time spent on the open arms as well as excursions into the end of the open arms and reduced scanning, indicative of an anxiogenic-like action. Injection of 5 and 250 mg/kg SP had no significant effect on the behavior in the maze. In the second experiment, SP was microinjected unilaterally into the NBM region in 3 doses ranging between 100 pg and 100 ng. Immediately after injection, the rats were tested on the EPM for 5 min followed by a 5 min test trial on the SIT. For rats treated with SP at 1 ng, an anxiolytic effect was observed in that these animals spent more time on the open arms of the EPM and showed an increase of time spent in social interaction (sniffing, following, grooming, crawling); intrabasalis SP injections at doses of 100 pg and 100 ng did not influence rats' behavior on the EPM and SIT. Furthermore, the anxiolytic-like effects of SP were specific in that the inverse sequence of the SP-molecule did not influence the behavior of the rats on the anxiety tests used. In sum, these results show that SP can have anxiolytic-like properties in addition to its known memory-promoting and reinforcing effects, supporting the hypothesis of a relationship between anxiety, memory and reinforcement processes.

Poster abstract

## LM/8

GRIP FORCE CONTROL DURING MICROGRAVITY. *J. Herms-dörfer[1], C. Marquardt[1], A. Zierdt, J. Philipp[2], D. Nowak[2], N. Mai[2]. [1] Clinical Neuropsychology Research Group (EKN), Krankenhaus München-Bogenhausen, Dachauerstr. 164, 80992 München, Germany. [2] Department of Neurology, Ludwig-Maximilians-Universität München*

Studies on healthy subjects have shown that grip forces produced during manipulation of objects are precisely controlled according to external requirements. For example, the grip force is precisely adjusted to the object's weight and surface structure during lifting; inertial loads arising when grasped objects are accelerated are compensated for by simultaneous changes in the grip force (Flanagan and Wing, 1995). Patients with brain lesions have been reported to have several disturbances of grip force control. We studied the reaction of normal subjects to rather drastic perturbations of object properties induced by changes of gravity during parabolic flights.
Two subjects held a manipulandum with a mass of 0,5 kg and equipped with grip force and acceleration sensors during successive parabolas of a parabolic flight. Subjects were instructed to hold the manipulandum stationary or to move it in a predefined manner during the parabolas, which consisted of 20 s long 0-G phases (weightlessness) preceded and followed by 2-G phases lasting about 10 s.
The weight of the object was doubled during the 2-G phase and was zero during the 0-G phase. Force recordings in the static condition showed that the subjects precisely adjusted their grip force to the weight changes and produced a low force of only 2 N during weightlessness. In the dynamic condition the object's weight was again zero during the 0-G phase, but inertial loads arose due to the acceleration of the mass. During up and down movements (frequency 2 Hz, amplitude 30 cm) in 1- and 2-G phases the external load was maximal at the lower turning point and minimal at the upper turning point due to the vectorial summation of weight and direction- dependent inertia. Without the contribution of weight during the 0-G phase, however, load had a maximum at the lower **and** upper turning points. The recordings again exhibited a near perfect adaptation of grip forces to these rather drastic changes of load, despite the fact that up and down movements did not change in speed or amplitude.
The results showed that the central nervous system is capable of taking into account changing gravity during the continuous calculation of grip forces. Thus, highly economical grip force control was maintained by the subjects despite their having no experience of microgravity before the parabolic flights.
*References:* Flanagan JR, and Wing AM (1995) The Stability of Precision Grip Forces During Cyclic Arm Movements with a Hand Held Load. Exp Brain Res 105: 455–464

## LM/9

BEHAVIOURAL EVIDENCE FOR FUNCTIONAL CONVERGENCE OF INTRINSIC AND EXTRINSIC SPATIAL INFORMATION IN BATS. *P. Höller; P. Krasemann; U. Schmidt. Zoological Institute, University of Bonn, Poppelsdorfer Schloss, D-53115 Bonn, Germany*

In addition to their excellent exteroceptive senses, bats are able to navigate by means of idiothetic orientation (the concept of idiothetic orientation has been introduced by Mittelstaedt & Mittelstaedt [Fortschr. Zool. 21, 1973] to describe the ability of animals to orient themselves without any exogenous spatial information). In accord-ance with this finding it has been observed several times that bats, when flying in a well known flight space, collide with obstacles which haven't been present before, or – the other way around – they performed evading manoeuvres although an obstacle had been removed. The presented work was designed to investigate the relative importance of idiothetic and allothetic (i.e. the opposite concept to idiothetic) orientation systems in bats. Exemplarily the orientation behaviour of the neotropical bat *Phyllostomus discolor* (Phyllosto-midae) during a straight ahead flight was examined. The animals were trained to proceed directly from a defined starting point to a defined landing site, which was marked by a landmark on either side. In order to allow multimodal landmark orientation, one mark was represented by a yellow LED while the other was a solid plastic block (20×5×3 cm, for echo-acoustic perception). For an investigation of the relative importance of landmarks and dead reckoning respectively, series of standard flights were intermitted by flights with shifted landmarks (following referred to as exp. 'LM'), shifted starting alignment of the bats (exp. 'SA'), and combined shifting of both landmarks and starting alignment (exp. 'B'). In all critical experiments the bats could be rewarded additionally at an alternative landing site which was located between the shifted landmarks (exp. 'LM') resp. in straight direction of the shifted starting alignment (exp. 'SA'). The distance between the landmarks as well as the relative position to each other (LED:left side, block: right side) was kept always the same. Computer based investigations of the trajectories in experiment 'LM' proved that the bats used the landmarks even after more than 300 flights in order to orient themselves. Their routes were shifted towards the direction of the alternative target (allothetic component of orientation). Remarkably, none of the bats hit the alternative target but they landed between the alternative and the standard target location. In experiment 'SA' several parameters of the trajectories revealed that the shifted starting alignment also affected the bats' choice of the landing area (idiothetic component of orientation). Nevertheless all bats started correcting manoeuvres shortly after their take off resulting in a heading towards an area between the standard and the alternative landing site. Functional convergence between idiothetic and allothetic orientation became evident in experiment 'B', when both landmarks and starting alignment were shifted. The effects of both experiments 'LM' and 'SA' were superposed resulting in a heading of the bats towards the alternative landing site.
Supported by the Deutsche Forschungsgemeinschaft (Schm 322/15-1)

## LM/10

IMPLICIT KNOWLEDGE OF INFORMATION PRESENTED DURING GENERAL ANAESTHESIA. *C. Hübner, D. Schwender, E. Pöppel: Institute of Medical Psychology; LMU Munich; Goethe-strasse 31; D-80336 München*

This study deals with the phenomena of implicit knowledge, in particular with modalities of non-conscious information processing, such as unconscious perception and cognition in altered states of waking consciousness (ASC) as for example cognition of intra-anaesthetic stimuli (that is cognition of information applied during general anaesthesia).
The initial hypothesis of this study has been that there is a referential system for phenomena of implicit knowledge with two orthogonal dimensions: 1. The modality of cognition (that is conscious or non-conscious) and 2. The state of consciousness (normal/waking state of consciousness and ASC).
The central question aimed at in this study was to examine whether there can be validated an effect, controlling behaviour due to non-consciously perceived information and due to information perceived in the ASC caused by general anaesthesia.
Further we looked for answers concerning potential differences of implicit knowledge evoked by differences in the intensity of stimuli presentation (loudness).
Another aim of the experimental design was to show a priorisation of the non-verbal method of getting access to implicit knowledge compared to the verbal one.

In a blind, controlled investigation we studied 150 patients (ASA 1 or 2) undergoing general or regional (local) anaesthesia at the Klinikum Innenstadt of the Ludwig Maximilians Universität in Munich. During anaesthesia the patients heard a 10-minute-text-tape with positive suggestions, either sub- or supraliminally (according to the allocation to the group). The controlling group was set up, so that the patients of this group did not get any textual information. 1 to 5 hours after operation all patients were asked about their intra-anaesthetic experiences.

The methods used to get access to implicit knowledge were both: verbal and non-verbal.

The verbal access was accomplished by questioning and for the non-verbal access the patients were asked to choose those picture cards, which they associate with their intra-anaesthetic experiences. The cards had to be chosen out of a prearranged set of picture cards. The set of picture cards contained such cards, which had a strong connection to the presented text as well as cards which obviously were far away from being associated with the presented text.

The main results are as follows:

1. Cognition in ASC, in particular during general anaesthesia, cannot simply be disregarded. The results show a significant difference between experimental and controlling group with p < 0,001 for the non-verbal access method.
2. Cognition in ASC cannot be functionally equalled with non-conscious cognition. The results show significant group differences with p = 0,04 for the non-verbal access method.
3. The results validate the experimental paradigm concerning the priorisation of the non-verbal method of getting access to implicit knowledge of intra-anaesthetic information compared to the verbal one.
4. The data suggest that there may exist differences in the information processing of stimuli presented in the ASC caused by general anaesthesia between those applied with low and those applied with high intensity (loudness of the acoustic stimuli).

## LM/11
DISORDERS OF MEMORY FUNCTIONS AND THE HIPPOCAMPAL FORMATION: FINDINGS IN PATIENTS WITH WELL-DEFINED UNILATERAL LESIONS. B.O. Hütter[1], K. Niemann[2], V.A. Coenen[1], J.M. Gilsbach[1]. Departments of [1] Neurosurgery and [2] Neuroanatomy, Medical Faculty of the University of Technology (RWTH), Pauwelsstr. 30, 52057 Aachen, FRG

Objective: The hippocampal formation is a complex group of anatomical structures. Among neuropsychologists, there is a substantial disagreement about the significance of the hippocampal formation for memory functions. Many of the cases presented in the literature have lesions which are not restricted to the hippocampal formation or were only supposed as in the cases with hypoxic brain damage. Furthermore, most lesions were not analyzed by modern MRI techniques. Therefore, we applied a measure for digital 3D analysis of MRI data in order to define exactly the anatomical structures involved in patients with hippocampal lesions.

Patients and methods: Three patients could be studied who were selected for unilateral, small (< 1 cm) and well-defined lesions in the hippocampal formation of comparable etiology. All patients were female and strongly right-handed. For brain mapping we pursued a strategy orientated to Boole's logic. We permitted only one "yes" relation to only one exclusive anatomical structure in order to enable simple and definitive attributions of functions to a given brain lesion of interest. Neuroanatomical localisation was performed by means of a software package (SulcusEditor) analyzing MRI data. The SulcusEditor enables the digital editing the topography of deep-seated processes by pursuing the sulcal morphology. First, the collateral sulcus was identified and edited enabling the identification and the labeling of associated anatomical structures in the neighbourhood. The MRI sequences (FLASH 3D; T1-weighted;1.6 mm; TI = 40; TR = 0.04) were edited interactively on a SUN/SPARC workstation. All patients were submitted to a neuropsychological examination including tests for aphasia, several aspects of attention and memory functions. A cognitive deficit was defined as a test score two standard deviations or more below the population mean according to the test norms.

Results: Table 1 gives the anatomical structure involved, the histology of the process and the memory impairments found.

Conclusions: By means of the SulcusEditor, the lesions studied could be attributed reliably to a single defined anatomical structure. The neuropsychological results suggest that only damage to the uncinate fascicle is associated with the drastic memory disorders frequently attributed to the hippocampus. However, we found also impairments in functions of attention and language suggesting that the hippocampal formation is not only involved in memory processing. The segmentation of the relatively small temporo-basal structures was diffcult. Segmentation and volume-rendering algorithms should be improved in order to identify and visualize even small temporo-basal structures.

## LM/12
SPRAGUE-DAWLEY RATS DO NOT USE SPATIAL ALLOCENTRIC STRATEGIES IN A MORRIS WATER MAZE TASK. J. Fey, J. L. Martinez, Jr.,and E.J. Barea-Rodriguez. University of Texas at San Antonio, Division of Life Sciences, 6900 N. Loop 1604 West, San Antonio, TX 78249

The Morris water maze is used extensively to investigate the neurobiology of learning and memory using rats (Morris et al., 1986) and mice (Silva et al., 1992). Animals may use two types of spatial learning strategies to find a hidden platform, either egocentric or allocentric. Using an allocentric strategy, navigation to find a hidden platform is guided by extramaze cues which are visible from the pool. Because the location of the platform does not change over trials, it is assumed that rats learn to associate the location of the platform with the extramaze cues (Mojdeh and Bures, 1996). Using an egocentric strategy the animal does not use extramaze cues to find the hidden platform. Rather the subject forms a spatial relationship between itself and the platform or goal (Mojdeh and Bures, 1996). We were interested in investigating whether albino Sprague-Dawley rats, which are commonly used in this task, use allocentric or egocentric strategies to learn the location of the hidden platform in the Morris water maze. The maze was surrounded by curtains and salients cues were attached to compass point locations on the curtains. The rats were trained to find the platform for 6 consecutive days. On Day 7, the rats were divided into three different groups; CONTROL, NO CUES (salient visual cues are removed), and REVERSAL (the platform location is reversed). In the probe trials we find that no experimental manipulation impaired spatial perfor-

Table 1

| Patient | Histology | Structure involved | Verbal STM1 | Figural STM[1] | Verbal LTM[2] | Figural LTM[2] | Spatial memory | Faces recognition | Retrograde Amnesia |
|---------|-----------|-------------------|-------------|----------------|---------------|----------------|----------------|-------------------|--------------------|
| F.R | PNET (WHO I) | parahippocampal gyrus right | No | Yes | Yes | No | No | No | No |
| S.B. | Cavernoma | alveus right | No | Yes | No | No | Yes | No | No |
| B.U. | Glioma (WHO II) | uncinate fascicle right | Yes | Yes | Yes | Yes | Yes | No | Yes |

[1] short-term memory; [2] long-term memory

mance either on Day 7 or Day 8. These findings suggest that under the experimental conditions used here albino Sprague-Dawley rats did not use distal visual cues to solve the Morris water maze problem and likely used an egocentric strategy.

Supported by NIH Grants GMO7717 (MARC) , DA 04195 (JLM), and GM08194-17S1 and Faculty Research Award (EJBR)

## LM/13
MOTOR PERFORMANCE OF ACTIONS: A MEMORY ENHANCING STRATEGY IN PATIENTS WITH BRAIN DAMAGE. *Günter Kriz. EKN, Clinical Neuropsychology Research Group, City Hospital Bogenhausen, 80992 München, Dachauer Str.164, Germany*

In healthy persons the recall of enacted action phrases (e.g. "throw a stone") is superior to recall of action phrases without enactment 1,2. In the rehabilitation of patients with memory disorders following brain injury some important questions arise concerning this enactment effect.

The purpose of this study was to examine if the motor encoding of action events could be a suitable strategy for patients with memory deficits to enhance memory performance, and under which circumstances this strategy promotes the highest proficiency. Another goal of the study was to investigate which mechanisms contribute to the enactment effect and how they are related to common verbal memory processes.

Two groups of patients participated in this study. Patients with organic amnesia but no motor disorders (n = 18) and patients with motor disorders but no cognitive disorders. Patients with moderate motor disorders (n = 22) and patients with severe motor disorders (n = 16) were analysed separately.

Members of all subject groups were assigned to four encoding conditions.
1. Verbal instruction (the unspecific instruction to hear the action phrases with the remark that they will have to be recalled later)
2. Visual imaginal instruction (the instruction to imagine another person who is performing the actions)
3. Motor imaginal instruction (the instruction to perform the actions mentally)
4. Motor enactment instruction (the instruction to perform the actions symbolically)

The analysis of the results of the patients with organic amnesia reveals a significant main effect for encoding condition. Recall in the visual imaginal and in the enactment condition was superior to the verbal condition. Recall in the motor imaginal condition however did not differ from recall in the verbal condition.

The pattern of results in patients with moderate motor disorders is quite similar, despite the fact that the list of action phrases in these patients included more items than the itemlist in patients with memory disorders.

In patients with severe motor disorders the results are clearly different. These patients show no enhancement of memory in the conditions motor imagination and motor enactment, compared to the verbal condition. The visual imaginal condition however leads to a higher recall of action phrases than the verbal condition.

In conclusion these data show that motor enactment is an effective strategy to enhance memory for action phrases in patients with organic amnesia. Moderate motor disorders have not influence on this result. Severe motor disorders, however, have a clear impact on motor encoding and suppress the enactment effect.

In contrast to motor enactment, motor imagination shows no effect on verbal memory for action phrases. Thus, for the investigation of the enactment effect it seems useful to distinguish different components of motor actions. In view of the present results, the actual and visible component of motor action is an important prerequisite for the enactment effect.

## References
1. Cohen, R. L. (1989). Memory for action events: The power of enactment. Educational Psychology Review, 1(1), 57–80
2. Engelkamp, J. (1995). Visual imagery and enactment of actions in memory. British Journal of Psychology, 86, 227–240

## LM/14
SHORT-TERM MEMORY FOR STIMULUS DURATIONS IN RATS: EFFECT OF AGE. *Pascale Leblanc and Monique Soffié. Psychobiology unit – University of Louvain – 10 place du Cardinal Mercier – 1348 Louvain-la-Neuve – Belgium*

The effect of age on rats' memory for event duration was investigated in two experiments using a symbolic delayed matching-to-sample procedure. In experiment I, four age groups (6, 12, 18 and 24 months) of rats were trained to discriminate a short (2 sec) and a long (10 sec) signal durations. When a retention delay of variable length (1, 2, 3 and 4 sec) was introduced between sample and comparison stimuli, rats responded as though long sample had been short. This choose-short effect already appeared for a retention delay of 2 sec in rats aged 12, 18 and 24 months, and only after a retention delay of 4 sec in the youngest (6 months old) animals. In order to test whether a long sample duration becomes subjectively shorter during the retention sessions, a more sensitive test (Experiment II) involving signals that are close to the borderline between the short and long categories was used. This experiment, carried out only in young (6 months) and old (18 months) rats, consisted to determine the psychophysical function relating probability of a long response to signal duration and to determine how a retention interval affects this function. The question was whether or not the point of subjective equality (PSE: i.e. the duration that the animal equally often classifies as long or short) increased in both age groups when the retention delay increased. Results show that the value of the PSE was significantly increased in 6 months old rats for all retention delays (1 to 4 sec). In 18 months old rats, the psychophysical function was displaced horizontally to the right after all retention delays but the increase in the PSE was never significant. These results suggest, contrary to Church (Learning and Motivation, 1980, 11: 208–219), that in young rats sample durations could be coded analogically and that the forgetting of a signal duration during retention intervals occured on a time dimension which provides support for the subjective-shortening hypothesis. In old rats there was no such evidence that the forgetting of a signal duration occured on the time dimension.

*This research has been supported by an «Action Concertée» grant n° 92/97-158 from the Ministry of the Belgian French Community.*

## LM/15
MITOCHONDRIAL DNA EFFECTS ON NEUROBEHAVIOURAL CHARACTERISTICS IN TWO INBRED MOUSE STRAINS: *Lalla Fatima Maarouf, F. Sluyter, C.C.G. Marican, W.E. Crusio and P.L. Roubertoux: UPR 9074 CNRS, Génétique, Neurogénétique et Comportement, Institut de Transgénose, 3B rue de la Férollerie, 45071 – Orléans, France*

Mutational damage to human mitochondrial DNA (mtDNA) may lead to neurodegenerative disorders. Genetically well defined animal models may shed some light on the mechanisms underlying these disorders. Towards this end, reciprocal congenic lines for their mtDNA were developed from the highly inbred NZB (abbreviated N) and CBA/H (abbreviated H) strains, which differ in their origin of mtDNA (N: *M. Musculus brevirostris*; H: *M. Musculus musculus*). The resultant congenic lines are designed $H^{mtN}$ (H background with N mtDNA) and $N^{mtH}$ (N background with H mtDNA). These congenic lines only differ from their parentals with respect to their mtDNA. Therefore, differences between a congenic line and its parental can only be attributed to that part of the genome for which they are congenic (i.e. mtDNA). Both parental (H, N) and congenic ($H^{mtN}$, $N^{mtH}$) lines were examined for 5 days in an 8-arm radial maze, a paradigm known to test spatial learning. Behavioural measures were total numbers of errors and mean number of correct entries in the first eight arms sampled on days 3–5. Per genotype, 12–16 animals were used. The results show an effect of the origin of both the genetic background and the mtDNA with the H background and H mtDNA increasing spatial learning. In addition, the sizes of the intra- and infrapyramidal mossy fibre (IIPMF) terminal fields, known to correlate with spatial memory, are being quantified. We conclude that, at least for these two strains, variation in mtDNA strongly affects neurobehavioural characteristics. Therefore, this set of parental and reciprocal congenic lines may be a useful animal model for studying mtDNA-dependent human neurodegenerative syndromes.

## LM/16
BEHAVIORAL EVALUATION OF TS65D MICE: FOCUS ON REFERENCE AND SPATIAL WORKING MEMORY DEFICITS.
C. Martínez-Cué[1], I.F. Vallina[1], C. Baamonde[1], R.M. Escorihuela[2], A. Fernández-Teruel[2], M. Dierssen[1], A. Tobeña[2], J. Flórez[1]. [1] Dept. Physiology and Pharmacology, School of Medicine, Univ. of Cantabria, 39011 Santander; [2] Medical Psychology Unit, School of Medicine, Autonomous Univ. of Barcelona, 08193 Bellaterra, Barcelona, Spain

Ts65Dn male mice (TS) and their control littermates (CO) were evaluated in test measuring sensorimotor reflexes, exploration, locomotor activity, emotional reactivity and spatial learning. No deficits appeared in TS compared to CO mice in visual acuity reflex, prehensile reflex, traction capacity, motor coordination and equilibrium. By contrast, head dipping behavior in the hole board was increased in TS mice with respect to the CO group, showing a higher repetition rate of previously explored holes. Crossings in the open field and total arm entries in the plus maze were higher in TS than in the CO group during the dark phase of the light-dark cycle. Entries into the open arms of the plus maze were increased overall in TS mice compared to CO mice, but no differences were found in time spent in the open arms. TS mice showed higher scape latencies and longer path lengths than CO mice in a repeated acquisition paradigm in the Morris water maze. This performance deficit was more marked in the second trials than in the first trials of several trial pairs. Thus, TS mice do not show impaired sensorimotor performance, they are hyperactive under certain conditions and present a spatial learning impairment which appears to be more pronounced in working than in reference memory. (Supported by Ramón Areces and Marcelino Botín Foundations, FISS 95/1779 and DGCYT PC94-1063. CMC and IFV have predoctoral fellowships from Gobierno Vasco and Real Patronato para la Prevención y Atención a Personas con Discapacidad, respectively).

## LM/17
SPRAGUE-DAWLEY RATS DO NOT USE SPATIAL ALLOCENTRIC STRATEGIES IN A MORRIS WATER MAZE TASK.
J. Fey, J. L. Martinez, Jr.,and E.J. Barea-Rodriguez. University of Texas at San Antonio, Division of Life Sciences, 6900 N. Loop 1604 West. San Antonio, TX 78249

The Morris water maze is used extensively to investigate the neurobiology of learning and memory using rats (Morris et al., 1986) and mice (Silva et al., 1992). Animals may use two types of spatial learning strategies to find a hidden platform, either egocentric or allocentric. Using an allocentric strategy, navigation to find a hidden platform is guided by extramaze cues which are visible from the pool. Because the location of the platform does not change over trials, it is assumed that rats learn to associate the location of the platform with the extramaze cues (Mojdeh and Bures, 1996). Using an egocentric strategy the animal does not use extramaze cues to find the hidden platform. Rather the subject forms a spatial relationship between itself and the platform or goal (Mojdeh and Bures, 1996). We were interested in investigating whether albino Sprague-Dawley rats, which are commonly used in this task, use allocentric or egocentric strategies to learn the location of the hidden platform in the Morris water maze. The maze was surrounded by curtains and

salients cues were attached to compass point locations on the curtains. The rats were trained to find the platform for 6 consecutive days. On Day 7, the rats were divided into three different groups; CONTROL, NO CUES (salient visual cues are removed), and REVERSAL (the platform location is reversed). In the probe trials we find that no experimental manipulation impaired spatial performance either on Day 7 or Day 8. These findings suggest that under the experimental conditions used here albino Sprague-Dawley rats did not use distal visual cues to solve the Morris water maze problem and likely used an egocentric strategy.
Supported by NIH Grants GMO7717 (MARC) , DA 04195 (JLM), and GM08194-17S1 and Faculty Research Award (EJBR)

## LM/18
DOPAMINERGIC MODULATION OF VISUAL WORKING MEMORY IN MAN. U. Müller. Max-Planck-Institute of Cognitive Neuroscience, Inselstr. 22, D-04103 Leipzig

Studies with single-cell recording in monkeys and functional neuroimaging methods in man have shown that the prefrontal cortex plays an important role in mediating working memory, i.e. the short-term maintainance of informations that are necessary for on-going decision-making (Fuster 1995, Goldman-Rakic 1996). Studies with intracerebral application of dopaminergic drugs claim a particular role of the D1 receptor subtype for the modulation of delay-related neuronal activity within the prefrontal cortex (Sawaguchi et al. 1994; Williams & Goldman-Rakic 1995). This can be explained by the tenfold higher cortical density of D1 receptors as compared to D2 receptors (Hall et al. 1994; Goldman-Rakic et al. 1997). In order to further investigate the "dopamine link" (Desimone 1995) of working memory in man and to look for differential D1 versus D2 receptor contributions we assessed the effects of pergolide, a mixed D1/D2 receptor agonist, and bromocriptine, a specific D2 receptor agonist, on a visual delayed matching task.
16 volunteers (8 female), age 19 to 29 (23.8 ± 2.9), received either 0.1 mg (per 80 kg body weight) of pergolide or identical placebo capsules on two seperate days in a double-blind, balanced design. Another 16 volunteers (8 female), age 19 to 29 (23.1 ± 3.0), received 2.5 mg (per 80 kg body weight) of bromocriptine or identical placebo tablets. A pretreatment with $3 \times 10$ mg of domperidone, a peripherally active D2 antagonist, was performed in both groups to reduce side-effects. The study was approved by the ethical committee of the regional board of physicians. Our working memory paradigm was a visuo-spatial delayed matching task implemented on a PC. The subjects had to memorize the location of a random-generated 7-point pattern and to compare it after 2, 8 or 16 sec with a second pattern that was either equal or slightly shifted within the frame. The task was designed with the intention to present unique stimuli at each trial and to require minimal motor demands. The main task consisted of 180 balanced trials and lasted about 50 minutes.
As predicted from pilot studies the paradigm showed significant error and reaction time (RT) increases with longer delays in both studies ($p < 0.001$). Pergolide, but not bromocriptine, reduced the error rates in the 16 sec delay condition and there was a significant drug × delay interaction in the pergolide study ($p < 0.05$). RT analysis revealed no speed-accuracy trade-off. There was also a significant effect of pergolide on two mood ratings (STAI-1 and Bf-S), i.e. subjects felt worse at the end of the pergolide session but not so in the corresponding placebo session. There was, however, no significant correlation between the memory and mood parameters. No significant effects of the two dopamine agonists on two paper-pencil tests of attention were seen. Delay-dependent errors and RTs, of the placebo days were not statistically different between the two groups.
Only pergolide, a mixed D1/D2 agonist, but not bromocriptine, a selective D2 agonist, reduced delay-dependend errors in a visuo-spatial working memory task when comparable doses (Wachtel 1991) were given to two balanced groups of volunteers. This findings are in accordance with the monkey literature (Arnsten et al. 1994) and support a preferential role of the D1 receptor in working memory modulation.

## LM/19

**DIFFERENTIAL CONDITIONING OF AUTONOMIC RESPONSES TO LATERALISED PREATTENTIVE EMOTIONAL STIMULI IN PATIENTS WITH MEDIAL TEMPORAL LOBE LESIONS.** *M. Peper[1], S. Karcher[1], J. Saar[2], R. Wohlfarth[2], P. Martin[2], G. Reinshagen[2]. [1] Research Program in Neuropsychology/Neurolinguistics and Department of Psychology, University of Freiburg, 79085 Freiburg, Germany, [2] Südwestdeutsches Epilepsiezentrum, 77694 Kehl, Germany*

Recent neurobehavioral research indicates that the medial temporal lobe, in particular the amygdala, is involved in aversive conditioning (e.g., LeDoux, 1992). However, the effects of medial temporal lesions on aversive associative learning are not well investigated in humans (see Bechara et al., 1996, for case studies). Psychophysiological findings suggests that hemisphere asymmetries must be considered, when the encoding or recall of conditioned electrodermal responses to preattentive conditioned stimuli (CS) is investigated (Johnsen & Hugdahl, 1991, 1993). The objective of the present study was to investigate the mechanisms of emotional associative learning by applying a visual half field paradigm with lateralised CS-presentations to patients with unilateral selective amygdalohippocampectomies (AHE). We hypothesise that the left visual field resistance-to-extinction effect for preattentive negative CS+ is reduced following right AHEs, i.e. contralesional presentations of the CS+ are expected to be associated with a deficit of conditioned responding.

We investigated patients with AHEs of the left (N = 14) and right (N = 12) temporal lobe, as well as a sample of N = 13 healthy volunteers matched for sex and age. A differential conditioning paradigm with emotional expressions (negative (CS+) and positive (CS−) Ekman faces) was applied. The US was an aversive vocalization (3 s, 95 dB). CS were presented preattentively during the habituation and extinction phases using a backward masking procedure. CS identification was determined by way of a perceptual threshold screening separately for each participant and for each experimental condition (SOAs were between 10 to 80 ms). Physiological indicators of conditioning were bilateral EDA and orbicularis oculi EMG, heart rate acceleration and vasoconstriction.

A group x visual half field x experimental phase effect indicated a differential autonomic responding of the investigated groups. A resistance-to-extinction effect for masked negative facial expressions presented to the left visual field was found for the electrodermal first interval response and for heart rate in controls. In AHE patients, electrodermal orienting appeared to be normal and discrimination of neutral faces was not significantly impaired. However, AHE patients did not show evidence for a delayed extinction of autonomic responses. In particular, right AHE patients did not respond to the negative CS+, but rather to the positive CS− presented to the lesioned hemisphere. Further analyses indicated that the reduced autonomic reactivity to the CS+ was a consequence of an impaired *acquisition*. This observation corresponds to findings implicating the amygdala in the *encoding* of aversive meaning. Since stimuli were presented *with* awareness during the acquisition phase, the interaction of the amygdala with frontal regions via its striatal connections might be important for the conceptual decoding of affiliative emotional features of the CS-US association (funded by the Deutsche Forschungsgemeinschaft, Pe 499/2-1, 2-2).

## LM/20

**THE PATH INTEGRATION PERFORMANCE OF MICE.** *L. Pörtner; P. Höller; U. Schmidt. Zoological Institute, University of Bonn, Poppelsdorfer Schloss, 53115 Bonn, Germany*

Path integration enables animals to find the way back to the point of departure after an outward journey, without the use of environmental orientation cues. For this purpose, subjects have to store all linear and rotational movements they have made during their outward run in order to reckon their present position and plan their return itinerary.

Two different experimental setups were used to measure the exactness of the path integration system of female mice and to find out whether mice would trust more in their assessment of distances or in their assessment of angles.

In experiment 1 female mice had to search for one of their sucklings, which had been removed from the litter and placed at the centre of a circular arena. When the dams had found their offspring, they picked it up and retrieved it to the nest, which was directly attached to the test ground (test setup according to Mittelstaedt and Mittelstaedt, 1980). Rotations of the arena before each trial guaranteed that the good homing performance of all test animals was not caused by the use of landmarks (mice need a couple of runs with stable landmarks in order to use them for orientation). In order to prove that the mice did not orient themselves according to orientation cues related directly to the nest entrance (e.g. ultrasonic vocalization of the remaining pups, scent of the litter) all trials were performed under infra-red light and standard trials were intermitted by critical trials where the whole arena was rotated by 90° or by 180° to either side, when the dams were just picking up their young at the centre of the arena.

In critical trials the dams always returned to the location, where the nest had been before the rotation. The directional error of the return paths never exceeded 6 degrees. There was no significant difference between trials with 90° or 180° of rotation.

In experiment 2 mice were trained under infra-red light to move from the periphery of a circular arena towards the centre of the testing ground in order to find a food item. When the animals had learned the task sufficiently, series of standard trials were intermitted by critical trials where the centred food item was replaced by three radially arranged food items in the periphery of the arena. In critical runs mice went out searching for food at the area where the reward had been in standard trials (in 91% of all critical trials the mice directly hit the target area, which was 1.5% of the whole arena surface). After being unsuccessful in finding any reward at the familiar feeding site the test animals started to search in an extended area. They all preferred to search for the removed food item in bows and circles which were directed towards the left and the right side. Search walks towards the food item which was placed straight ahead were very seldom. The centre of gravity of the search paths was located exactly at the familiar feeding site. We conclude that the path integration system of mice proves to be very exact and quite robust against passive rotations. Further we realized that the distance estimation turns out to be a stronger parameter of path integration than the angle assessment is.

## LM/21

**SEX DIFFERENCES AND OBJECT LOCATION MEMORY.** *Albert Postma & Edward de Haan. Psychological Laboratory, Utrecht University, Heidelberglaan 2, 3584 CS Utrecht, The Netherlands*

The existence of sex differences in cognitive abilities has long been acknowledged. Typically, males appear superior in spatial abilities, whereas females excel in verbal skills. Hormonal influences and differences in cerebral organization could be important.

There remains some controversy, though, on which specific tasks will show the differences and to what extent. In this vein, it was the purpose of the present study to establish whether sex effects exist in object location memory, and – if so – for which subcomponents. Object location memory requires subjects to reconstruct the positions of various previously studied objects. Recently (Postma & De Haan, 1996), it was argued that as such separate processes may be involved. First, one needs to remember the precise positions occupied (positional encoding per se). Second, one has to decide which object was at which position (object to position assignment). Third, both types of information need to be integrated in order to recall the exact positions of multiple different objects.

In the present study 20 males and 20 females (matched in age and education) were tested on the following task. A square frame (15×15 cm) was shown on a computer screen, containing always 10 objects, for 30 sec. Next the objects disappeared from the square to reappear on a row above it. Using the computer mouse subjects could relocate them. There were 3 relocation conditions. In the object-to-position-assignment condition, the positions where objects should be replaced were always premarked by a dot. In the positions-only condition, all objects were the same, so only the exact positions had to be retained. In the so called combined condition, one should place all different objects in the square, without any marking of the original positions. This condition thus comprised both positional encoding and object to position assignment. To establish the influence of strategic differences – females might rely more upon verbal processing of spatial information – a verbal interference task was introduced: while studying the stimuli subjects had to recite the word blah.

In the object-to-position-assignment condition no sex differences were found. In the positions only and the combined condition (ignoring in the latter the identities of the objects), however, a significant difference was obtained for positional reconstruction per se: males did better than females. In the combined condition also a second error measure was computed, taking into account the mean displacement between an object s original position and its reconstruction. Females performed as well as males here. It may further be noticed that, though verbal interference condition yielded main effects, there was no interaction with sex. Thus, the observed differences were not due to females relying more upon less efficient, verbal processing of spatial information.

In sum, the results demonstrate that males outmatch females in the encoding of the precise positional information, but not in the assignment of objects to positions. This relates to the distinction between categorical and coordinate encoding of spatial relations, and the presumed left hemispheric superiority for the former and right hemispheric advantage for the latter (Kosslyn, 1987).

Note. Part of these data were presented at the INS conference, Veldhoven 1996.

## LM/22

TEMPORAL CHARACTERISTICS OF CHICKS' (*Gallus gallus domesticus*) MEMORY FOR THE LOCATION OF A DISAPPEARED IMPRINTING OBJECT AND FOR THE SPATIAL SOLUTION OF A DETOUR TASK. *L. Regolin[1] and G. Vallortigara[1]. [1] Department of General Psychology, University of Padova, via Venezia 8, 35131 Padova – Italy. e-mail: regolin@psico.unipd.it. [2] Dipartimento di Scienze Filosofiche e Storico-Sociali, Laboratorio di Psicologia Sperimentale, Universita' di Udine, Via Antonini 8, 33100 Udine – Italy. e-mail: giorgio.vallortigara@ifp.uniud.it*

Previous studies analysed detour behaviour in chicks using an imprinting goal-object (a small red ball suspended by a thread) placed behind U-shaped barriers. In such conditions chicks can represent the goal and its spatial location even in the absence of any locally orienting sensory cues. We proceeded to investigate the temporal characteristics of the chick's representation of the no-longer perceived imprinting object, and the long-term retention of the solving strategy adopted. Chicks can easily learn to search for their imprinting object after having seen it disappear behind some obstacle. Memory of the imprinting object was studied in a "delayed reaction" test: five-day old chicks confined behind a glass partition watched the red ball disappear behind one of two screens. After a delay time (varying from 0 to 120 sec) chicks were released and could start looking for the ball. Each chick was given 16 consecutive trials, with the ball disappearing each time behind one of the two screens, randomly chosen. Chicks oriented their search towards the correct screen even with the longest time-interval, showing not only the ability to take into account and remember the directional cue provided by the ball movements, but also to be able to update this memory trial-by-trial, so that earlier trials were not interfering with the most recent one.

The nature of memory of the task itself was assessed in a separate set of experiments.Two-day-old chicks were faced with a spatial task requiring them to detour a U-shaped barrier in order to rejoin their companions placed beyond of it. Chicks were retested 30 minutes, 3 hours and 24 hours after a single successful detour had been performed at first trial. At retest, chicks took significantly less time to detour the barrier and maintained a good memory for the task even after a single 24 hours-long interval. Chicks that failed to solve the task at first trial took, 24 hours later, a time as long as that required by totally naive chicks. The effect cannot therefore be attributed uniquely to maturation, or to a lower emotional response to the novel environment due to previous exposure.

Chicks' representation of a disappeared object seems to last far longer and to be more detailed and sensitive to modifications in the actual situation than it was previously assumed. Memory of the task itself is likely to be of a different nature than memory for the object; solving the detour problem seemed to be necessary in order to form a long-lasting memory of it suggesting that an all-or-nothing process of learning is possibly involved in this task. The tecniques here employed lend themselves as promising tools for further investigation of the pharmacological and cellular properties of chicks' working- and long-term memory.

## LM/23

RECONSOLIDATION AFTER REACTIVATION OF MEMORY: ROLE OF b NORADRENERGIC RECEPTORS. *Pascal Roullet, Jean Przybyslawski & Susan J. Sara. Institut des Neurosciences, UPMC, 9 quai St Bernard, 75005, Paris, France*

Memory is reactivated by cues associated with the initial acquisition of information. Little is know about the physiological events underlying the retrieval processes involved, but we do know that the memory in its active form is labile and undergoes a post reactivation, time-dependent reconsolidation process, which is dependent upon NMDA receptors (Przybyslawski & Sara, 1997). The present study addressed the question of the role of b noradrenergic receptors in post reactivation reconsolidation in rats, it has been recently shown that the beta receptor antagonist, propranolol, can induce amnesia when administered to human subjects after learning information with high emotional content (Cahill et al, Nature, 371, 702). Rats are trained, 3 trials/day, in an 8 arm radial maze, to choose the same 3 baited arms, to a criterion of 3 consecutive trials with a maxium of one error per trial. The day after criterion is reached, the rat runs one trial to activate the memory and is injected with propranolol (10 mg/kg) or saline 5 min, 2 h or 5 h after the trial (independent groups). Twenty-four hours later, the rats are tested for retention with three trials, as during training. Propranolol injected at 5 min or 2 h, but not 5 h after the reactivation trial, induces amnesia, as measured by the number of errors on the three test trials.

To control for the specificity of action of propranolol on a reactivated memory, we performed a replication of the first experiment with the addition of a group which did not have the single reactivation trial, but received its injection in the animal vivarium. Rats were trained to the same criterion and were divided into reactivated and nonreactivated groups and were injected 2 h after the reactivation trial. When tested on the three trials, 24 h later, only those rats which had the reactivation treatment showed amnesia.

In both experiments, the amnesia was transient, with animals relearning the task within the test session, suggesting that the effect of the drug treatment was not proactive on performance. These results confirm the importance of b receptors in long term memory formation, up to two hours after acquistion and lend further support to the idea that reactivated memory is labile and triggers a reconsolidation process which may involve a recapitulation of the intracellular cascade involved in initial formation of long term memory.

Supported by CNRS (URA 1488) and European Neuroscience Programme (ESF)

## LM/24

MODULATION OF MEMORY, REINFORCEMENT AND FEAR PARAMETERS BY INTRA-AMYGDALA INJECTION OF CHO-LECYSTOKININ-FRAGMENTS BOC-CCK-4 AND CCK-8s: *S. Schildein; P. Gerhardt; C. Privou; H. Fink\*; R.U. Hasenöhrl; J.P. Huston: Institute of Physiological Psychology, University of Düsseldorf, Universitätsstr. 1, D-40225 Düsseldorf, Germany; \* Institute of Pharmacology and Toxicology, Medical Faculty Charité, Humboldt-University at Berlin, D-10098 Berlin, Germany*

This series of experiments examined the effects of the cholecystokinin (CCK) fragments Boc-CCK-4 and CCK-8s on memory, reinforcement and anxiety following unilateral injection into the central nucleus of the amygdala. In experiment 1, rats with chronically implanted cannulae were microinjected with different doses of CCK-8s (0.1, 1, 10, 100 ng) or Boc-CCK-4 (2, 20, 100 ng) and were tested on a one-trial uphill avoidance task. Post-trial injection of Boc-CCK-4 and CCK-8s was found to facilitate inhibitory avoidance learning in a dose-dependent manner. A single injection of 20 ng Boc-CCK-4 or 1 ng CCK-8s improved the retention performance, whereas the lower and higher doses had no effect. The hypermnestic effects of Boc-CCK-4 and CCK-8s were no longer evident when injection was performed 5 h, rather than immediately, after the learning trial, ruling out enduring proactive effects of the treatment on test performance. In experiment 2, the elevated plus-maze was used to gauge anxiogenic properties of intra-amygdala injections of Boc-CCK-4 and CCK-8s in memory-enhancing doses. The treatment with 20 ng Boc-CCK-4 and 1 ng CCK-8s did not influence the number of entries into and time spent on the open and enclosed arms of the maze as well as other anxiety-related behaviors, including scanning, risk-assessment and end-activity. In experiment 3, possible reinforcing effects of the CCK-fragments were examined. After intra-amygdala injection of Boc-CCK-4 or CCK-8s in memory-enhancing doses the rats were placed into one of four restricted quadrants of a circular open field (closed corral) for a single conditioning trial. Subsequent tests for conditioned corral preference in the open corral revealed no evidence for reinforcing or aversive effects of the CCK-fragments. These findings indicate that Boc-CCK-4 and CCK-8s facilitate memory processing upon injection into the central nucleus of the amygdala without exerting reinforcing or anxiogenous effects.
Poster abstract

## LM/25

DEVELOPMENT OF LEARNING AND MEMORY PROCESSES DURING POSTNATAL ONTOGENESIS AND THEIR DEPENDENCE OF GROWTH FACTORS IN MATURE RAT BRAIN. *V.V.Sherstnev, M.V. Pletnicov, Z.I. Storogeva, A.T.Proshin. P.K.Anokhin Institute of normal physiology RAMS B. Nikitskaya, 5,103009, Moscow, Russia*

Participation of growth factors in learning and memory processes in developing and mature brain was investigated. Short-term and long-term habituation of acoustic startle reaction (ASR) with simultaneous assesment of freezing responses in 18-, 30- and 90-days-old were studied. We have investigated effects of monoclonal antibodies (MAB) against associated with growrh of cerebellar and hippocampal cells protein A3G7, and effects of polyclonal antibodies (AB) against lectines R1 and CSL – factors of cell adhesion and recognition – on memory and learning in adult rats. It was found that 18-days old rats demonstrated only short-term habituation of ASR whereas 30- and 90-days old rats were able to elaborate as short- and long-term habituation. Fear conditioned freezing was observed in rats of all ages. Application of AB against R1 and CSL on cerebellar vermis has destroyed long-term habituation but not short-term habituation of ASR or conditioned freezing. Application of 5 mkg MAB against A3G7 (obtained from Institute of Medicine and Biologic Kibernetic SD RAMS) on cerebel-lar vermis selectively destroyed long-term habituation and conditioned freezing in adult rats. At the same time applicarion of 50 ng of MAB destroyed only long-term habituation and there was no effect on freezing. Thus, we have

obtained some data about heterochronic developmental forming of different kinds of defensive learning during ontogenesis and geterogenic involvement in these kinds of behavior some types of growth factors in mature brain. These results are experimental evidences of ideas about similar molecular basis of growth and development processes and the processes of learning and memory.

## LM/26

LATERALIZATION OF SPATIAL MEMORY TASKS IN THE DOMESTIC CHICK (*GALLUS GALLUS*). *L. Tommasi\*, G. Vallortigara. \* Dipartimento di Psicologia Generale, Universita' di Padova, Via Venezia 8, 35131, Padova, Italy. Dipartimento di Scienze Filosofiche e Storico-Sociali, Laboratorio di, Psicologia Sperimentale, Università di Udine, Via Antonini 8, 33100, Udine, Italy*

This study was aimed at searching for possible differences in performance in spatial memory tasks between binocular- and monocular-tested chicks. A group of chicks was daily trained to find some food hidden under sawdust, by ground-scratching in the central position of a square-shaped arena. Training started on day 8. On day 16, we compared performance of binocular, left- and right-eyed chicks. As already shown (Tommasi, Vallortigara & Zanforlin in press), binocular chicks were able to locate accurately the central position of the arena. Left- and right-eyed chicks also showed good spatial memory. When tested in an arena of the same shape but a larger area, binocular and left-eyed chicks displayed searching behaviour at two different distances from the wall of the arena, one corresponding to the correct distance (i.e. centre) in the smaller (training) arena, the other to the actual centre of the larger test arena. Right-eyed chicks, in contrast, displayed searching only at a distance from the walls of the arena corresponding to the distance in the smaller training arena. Another group of chicks was trained in the square-shaped arena with a conspicuous visual landmark (a red stick) located at the centre. Binocular chicks learnt easily to localize food using the landmark, and appeared capable to maintain their spatial memory when tested under monocular conditions both in presence and in the absence of the visual landmark. However, when the landmark was displaced in a corner of the arena, left-eyed chicks searched in the centre of the arena disregarding the presence of the landmark, while right-eyed chicks searched mostly around the landmark. Moreover, when two landmarks were simultaneously located in the arena, one in the correct central position and one near to a corner, left-eyed chicks searched more accurately in correspondence to the central landmark than right-eyed chicks.
Results suggest that spatial memory in the chick may be differently served by the left and the right eye and their associated neural structures.
Tommasi, L., Vallortigara, G. & Zanforlin, M. (1997): Young chickens learn to localize the centre of a spatial environment; *J. Comp. Physiol. A*, in press

## LM/27

DIFFERENCES IN WORKING MEMORY SPAN DURING AGING IN RATS. *Martine Van Waas and Monique Soffié. Psychobiology unit – University of Louvain – 10, Place du Cardinal Mercier – 1348 Louvain-la-Neuve – Belgium*

It has been often reported both in human and animal that working memory is impaired during aging. As far as working memory span is concerned, data obtained in human subjects have given precise value for the span decrement in old subject (review in Van der Linden and Hupet «Le Vieillissement cognitif», Paris, PUF, 1993: 37–85). The present experiments were carried out in order to precise whether similar impairment could be shown in old animal (rats). Young adult (6 months old) and old (23–24 months old) male Wistar rats were either tested in a complex cone-field apparatus (for details, see Van der Staay et al., Psychobiology, 1990, 18(3) : 305–311) either in an eight-arm radial maze. In the cone-field, the task consisted in finding 6 out of 16 food pellets put in 6 out of 16 cones, leaving from four different starting areas. The 6 baited cones were clustered in a corner of the field, according to four configuration patterns. Each rat was confronted with one of the four configurations, according to a random series. In the radial maze, the task consisted in visiting once all eight baited arms. In order to avoid any systematic strategy, an inter-trial confinement of 10 sec was applied. In both tasks, working memory errors were defined as revisiting either a baited cone or a baited arm already explored. The memory span, i.e. the number of different baited cones or different baited arms visited before the first working memory error was calculated for the last trials of the training (trials 91–100 for the cone-field and trials 41–50 for the radial maze). Results showed that the working memory span was lower in old than in young animals. However, the age-related difference was stronger (about 33%) in the radial maze than in the cone-field apparatus (about 15%). These results could be explained by the differences of the working memory complexity (e.g. spatial vs non spatial, difference in the number of items to maintain in working memory) between the two tasks.

This research has been supported by an «Action Concertée» grant n° 92-97-158 from the Ministry of the Belgian French Community.

## LM/28

ACTIVITY OF PRIMATE DOPAMINE NEURONS IN A DISCRIMINATION AND BLOCKING PARADIGM. *P. Waelti, J. Mirenowicz, W. Schultz. Institut de Physiologie, Universite de Fribourg, CH-1700 Fribourg, Switzerland*

Midbrain dopamine (DA) neurons are phasically activated by unpredicted primary rewards (appetitive US) and by conditioned, reward-predicting stimuli (appetitive CS). However, they are not activated by predicted rewards. During learning, a transfer of the dopamine response occurs from the US to the CS. According to associative learning theories, only unpredicted reinforcers contribute to learning. This is based on the blocking paradigm in which a new stimulus is added to a fully conditioned stimulus. The new stimulus will not acquire associative strength because the reward is already predicted by the conditioned stimulus. In order to formally assess the relationships of midbrain dopamine neurons to the unpredictability of rewards, we tested their activity in a blocking experiment.

In a first step, we tested the discriminative capacities of dopamine neurons. In a classically conditioned discrimination task, monkeys were presented with two randomly alternating pictures which were either followed by liquid reward (A+) or not (B−). Animals discriminated well between the two stimuli, as documented by their licking behavior. DA responses were stronger for rewarded stimuli as compared to non-rewarded stimuli, both in terms of percentage of neurons responding and magnitude of responses.

As dopamine neurons discriminated between rewarding and non-rewarding stimuli, we tested blocking. A new stimulus was simultaneously presented together with each previously experienced stimulus, thus producing AX+ and BY+. We then tested whether X and Y could be conditioned. Licking behavior showed that prior pairing of stimulus A+ with a reward blocked the conditioning of stimulus X. In contrast, the previously unconditioned stimulus B did not block conditioning of Y. The electrophysiological responses of DA neurons showed a comparable blocking effect. More DA neurons were activated with higher magnitudes by the non-blocked stimulus Y as compared to the blocked stimulus X.

In conclusion, blocked stimuli produce weaker response in DA neurons, probably because they have not acquired associative strength and do not serve as a reward predictors for the animal. These results are consistent with the hypothesis that DA neurons are sensitive to the predictability of primary rewards. This would suggest that DA neurons code an error in the prediction of reward. Their response to unpredicted rewards would constitute an effective appetitive teaching signal.

## LM/29

MONKEY AND HUMAN: LEARNING FROM THEIR PUPILS. *L. Weiskrantz, A. Cowey, and C. LeMare. Dept. of Experimental Psychology, University of Oxford, UK*

The pupil of the normal human subject constricts in response not only to average increases in light energy, but also selectively to the spatial structure of a visual stimulus even when there are no energy changes. This enables one to measure visual acuity and sensitivity as a function of spatial frequency. It is known that pupillometric measures of acuity correlate well with those determined psychophysically for normal human observers. This purpose of the present study was to measure pupillary changes with stimuli delivered to the "blind" hemifields of monkeys with unilateral V1 removal, and also that of a human subject (G.Y.) with putative V1 destruction. The results show that there are small but reliable pupillary changes to flux-equated gratings in the blind fields both in monkeys and human. The response profile of both species is very similar: it is narrowly tuned, with a peak at about 1 cycle per degree, and with a cut-off acuity of about 7 or 8 cycles per degree, a significant reduction compared to the intact hemifield. The result also maps well onto the psychophysically-determined spatiotemporal response profile to gratings in the blind field, as determined independently for G.Y. Thus, a narrowly tuned spatiotemporal visual channel remains in monkey and human in the blind field that does not require the integrity of V1. The pupil can be used as a method of determining its shape and sensitivity, and other properties of residual visual capacity.

# Language and Speech

## LS/1

LETTER POSITION ENCODING AND MAGNOCELLULAR VISUAL FUNCTION. *P.L. Cornelissen[1], P.C. Hansen[2], J.F. Stein[2]*. [1] *Psychology Dept., Ridley Building, Newcastle University, Newcastle-on-Tyne, NE1 7RU, UK.* [2] *Oxford University Physiology Dept., Parks Road, Oxford, OX1 3PT, UK*

Recent research has shown that reading disabled children find it unusually difficult to detect flickering or moving visual stimuli, consistent with impaired processing in the magnocellular visual stream. Yet, it remains controversial to suggest that reduced visual sensitivity of this kind might affect children's reading. We suggest that when children read, impaired magnocellular function may degrade information about where letters are positioned with respect to each other. We tested this hypothesis in three experiments.

First, we looked for an association between magnocellular function (assessed by a coherent motion detection task) and letter position encoding in 54 adult subjects. In experiment one, we used a lexical decision task in which we showed subjects strings of five letters for a period of 35 milliseconds. Each presentation was followed by a pattern mask. Half the letter strings were words and half were anagrams. We predicted that those individuals who encode letter position poorly (i.e. those who performed worst on the motion detection task and have poor magnocellular function) ought to 'unscramble' anagrams, and respond to them as though they were real words; this is what we found. In a second experiment on the same 54 subjects, we used a 'constituent letter priming task' to look for an association between magnocellular function and letter position encoding. Normally in this reaction time task, people respond quickest if both the prime and target letter-strings contain the same letters in the same positions (e.g. prime = qxta, target = QXTA). Reaction times are slowed if the positions of the letters are changed between prime and target (e.g. prime = tqax, target = QXTA). Reaction times are slower still if none of the prime letters appear in the target string (e.g. prime = nspv, target = QXTA). These priming effects have been interpreted as direct evidence for immediate and automatic encoding of letter position during visual word recognition. We found that these priming effect were smallest in those individuals with impaired magnocellular function, consistent with failure to encode letter position accurately.

Finally, in a third experiment, we predicted that degraded letter position information should cause children to make reading errors which contain sounds not represented in the printed word (e.g. misreading VOCATION as VOCTION, or PERSON as PRESON). We call these orthographically inconsistent nonsense errors 'letter' errors. To test this idea we assessed magnocellular function in 58 children by using the same coherent motion detection task. We then gave these children a single word reading task and found that the likelihood of them making 'letter' errors was best explained by independent contributions from motion detection (i.e. magnocellular function) and phonological awareness (assessed by a spoonerism task). This result held even when chronological age, reading ability, and IQ were controlled for.

Taken together, the findings from these three experiments suggest that impaired magnocellular visual function may degrade information about letter position which, in turn, may affect how children read.

## LS/2

NEUROANATOMICAL CORRELATES OF VERB PROCESSING: EVIDENCE FROM PATIENTS WITH SELECTIVE IMPAIRMENT AND SPARING OF VERBS. A. DANIELE, A.M. DI BETTA, V. FILIPPINI, M.C. SILVERI, G. GAINOTTI. *Institute of Neurology, Catholic University, Largo A. Gemelli, 8, Rome, I-00168, Italy*

Previous investigations in brain-damaged patients have suggested the hypothesis that, in the dominant hemisphere for language, neural systems located in the temporal lobe might play a critical role in the production and comprehension of nouns, while analogous systems in the posterior frontal regions might have a similar role in the production and comprehension of verbs (Daniele et al. 1992, J. Clin. Exp. Neuropsych. 14, 3:396; Daniele et al. 1994, Neuropsychologia 32,1325–1341). Similarly, it has been proposed that the left temporal lobe and the left frontal lobe are involved in the retrieval of nouns and verbs, respectively (Damasio & Tranel 1993, Proc. Natl. Acad. Sci. 90:4957–4960). These hypotheses, however, are still controversial, particularly as concerns the possible anatomical substrates of verb processing (De Renzi et al., 1995, Cortex 31:619–636, 1995).

The objective of our study was to further investigate the issue of the anatomical regions putatively involved in lexical processing of verbs. We report two brain-damaged patients who showed a selective impairment of verbs and a selective sparing of verbs, respectively. They underwent magnetic resonance imaging (MRI) and neuropsychological assessment, including lexical-semantic tasks on nouns and verbs matched for word frequency and length.

Patient 1 was affected by a progressive aphasic syndrome, with a three-year history of word-finding difficulty. MRI revealed small areas of increased signal intensity in the corona radiata frontalis, while single photon emission computed tomography showed a reduction of regional cerebral blood flow in the left frontal regions. On tasks of oral and written naming, Patient 1 showed a disproportionate impairment on verbs as compared to nouns. On tasks of auditory and visual comprehension he performed quite accurately on both nouns and verbs.

Patient 2 was a 26-year-old patient who suffered HSE with partial recovery. MRI four months post-onset showed extensive lesions in temporal and limbic structures most marked on the left, with some additional lesions in the anterior cingulate gyrus and the adjacent mesial frontal regions. Sixteen months post-onset, Patient 2 was disproportionately impaired in oral and written naming of nouns, with relative sparing of verbs. On word-picture matching tasks, she showed a moderate impairment in the comprehension of nouns, with relative sparing of verb comprehension.

In conclusion, Patient 1 showed a predominantly left frontal lobe damage and was disproportionately impaired in oral and written naming of verbs. Patient 2, with extensive lesions (mostly involving the temporal lobes) which spared the lateral frontal regions, showed a relative sparing of verb processing across various naming and comprehension tasks. These findings support the hypothesis that neural systems critically involved in the production and comprehension of verbs might be located outside the temporal lobe, possibly in lateral posterior frontal regions.

## LS/3

CONTROL OF SPEECH RATE AND RHYTHM IN PATIENTS WITH LEFT HEMISPHERE LESIONS. *Karin Deger and Wolfram Ziegler. EKN – Clinical Neuropsychology Research Group, Department of Neuropsychology, City Hospital Bogenhausen, Dachauerstr. 164, D – 80992 München*

The speech of patients with left brain lesions is often characterized by deviant speech rhythm (Gandour et al 1994). The underlying problem of suprasegmental speech timing remains still unclear. It can be a consequence of problems at the segment level or of a general timing deficit. Speech is a rhythmically organized behavior with the syllable as its fundamental unit. An influential theory for the timing control of rhythmically organized motor sequences is the proportionate "timing" model (Viviani & Terzuolo 1980). It predicts, that ratios between successive time intervals remain approximately constant, irrespective of the speed at which the movement as a whole is executed. Search for evidence of proportional timing led to conflicting opinions of this concept. (Lövqvist 1991).

The ratios of relative syllable durations were investigated in two experiments: (1) In a syllable reiteration (SR) task subjects had to imitate or generate a rhythmically structured stimulus. (2) In a word production (WP) task subjects had to speak a four syllabic word at different rates. The perception of time was investigated by interval discrimination and controlled by pitch perception. 8 patients with apraxia of speech (AOS), 9 patients with aphasia but no speech apraxia (NoAOS) and 30 normals participated.

In the SR task the normals and NoAOS patients held the target rhythm best when they spoke at normal rate. The AOS patients in contrast maintained the rhythm best at slow speaking rates. This might point to the fact that not the relative durations, but the absolute durations are the major problem for these patients.

In the WP task it appeared difficult, even for the healthy controls, to preserve a normal word accent when speaking at reduced rate. The NoAOS patients couldn't speak with normal word accent when changing the speech rate in any direction. The AOS patients showed an aberrant speech rhythm in each condition. Adapting their speaking rate to their articulatory and phonological capabilities, they were not able to speak in an automatized way.

The discrimination tasks gave no indication for a specific time perception deficit in the patient groups.

Proportionate timing, occurs only in skilled, automatized speech movements within a narrow velocity range and not under artificial conditions. Under the condition of speaking at a low rate it is difficult, even for the healthy controls, to preserve speech rhythm. This might have caused the deviant metrical structure in AOS. The increasing speech control prevents an automatized proceeding of the movement and produces a break down of the speech rhythm.

These observations do not support the hypothesis that deficient time structures "on the surface", i.e. a deviant speech rhythm, are a result of a general timing deficit in the definition of a defective "internal clock".

**References**

Gandour J, Dechongkit S, Ponglorpisit S, Khunadorn F (1994) speech timing at the sentence level in Thai after unilateral brain damage. Brain and Language 46: 419–438

Löfqvist A (1991) Proportional timing in speech motor control. Journal of Phonetics 19: 343–350

Viviani P, Terzuolo C (1980) Space time invariance in learned motor skills. In: Stelmach GE, Requin J (Eds) Tutorials in motor behavior. North Holland Publishing Company, pp 525–533

## LS/4

LINGUISTIC FACTORS IN NEGLECT DYSLEXIA FOLLOWING RIGHT HEMISPHERE LESIONS – AN INTERACTIVE MODEL. *E.G. de Langen. Klinikum Passauer Wolf, Department of Neurology, Postfach 1263, D – 94083 Bad Griesbach, Germany*

Disorders of the visuo-spatial attention system cause neglect dyslexia in many patients with right hemisphere parietal lesions. Inconsistency in the results of reading single words lead to the question if omissions, misreadings and also correct responses are due more to physical parameters of the stimuli, like word length, or, if there may be some specific linguistic factors influencing reading performance in these patients.

Two patients, MD and JK, both with large parietal lesions after stroke but without left homonymous hemianopia and without any signs of aphasia were investigated with a large selection of single words (n = 540), which varied in length and showed different morphological and lexical compositions. Stimuli included monomorphemic words, words with a free and a bound morpheme, words containing a free morpheme embedded on the right side, compound words of two lexical morphemes and compound words of three lexical morphemes. The single words were presented in lists of 18 items each, printed in 6 × 4 mm black capital letters for reading aloud without time limit. The sheets were placed exactly in the midline of the patient's trunk.

The results showed a very good performance (87%/80% correct) for monomorphemic words and also for words with a free and a bound morpheme (70%/85% correct), fair to good results (62%/89% correct) for words containing a free morpheme embedded on the right side, poor performance for compound words of two lexical morphemes (38%/0% correct) and for compound words of three lexical morphemes (0%/0% correct). Word length was of no consequence for the level of performance. Error responses included mainly omissions of left side parts of compound words, left side omissions under morpheme level, resulting in real words after completion by one or more phonemes, self corrections and some neologisms.

A working model based on these data accounting for the influence of lexical status and morphological composition of words in reading of patients with neglect dyslexia is presented. This interactive model provides also explanations for the origin of error responses like omissions, additions, substitutions, self corrections and neologisms. It is assumed that attention is distributed as a gradient from left to right and, that this gradient may fluctuate in steepness individually, also as a result of linguistic factors. The atttentional deficit may arise at several stages of the information processing, the linguistic characteristics of the stimuli however may provoke top-down knowledge or cause bottom-up deficits on low-level stages. This model-guided interpretation of performance allows to locate different loci of breakdown and repair in the information processing system following deficient distribution of attention in patients with right hemisphere parietal lesions.

## LS/5

TOPOGRAPHICAL ANALYSIS OF EVENT-RELATED POTENTIALS IN LANGUAGE-RELATED TASKS IN APHASIC PATIENTS AND CONTROLS AS A METHOD TO INVESTIGATE CORTICAL REORGANIZATION. *C. Dobel[1], E. Zobel[1], B. Rockstroh[1], R. Cohen[1], P. Köbbel[2], P.W. Schönle[2]. [1] Department of Clinical Psychology, University of Konstanz, Fach M 500, 78457 Konstanz, Germany. [2] Kliniken Schmieder, Postfach 240, 78473 Allensbach, Germany*

Reorganization of cortical functions related to language production in aphasics were examined by means of the topographical pattern of event-related brain potentials (ERPs). Different experiments were designed to activate language-related processes (based on neuropsychological tests sensitive to dysfunctions in aphasia and cognitive models of language production) involving concept formation, word recognition, syntactic classification and phonological encoding. ERP's were recorded from 27 electrode locations on 5 acute aphasic patients with time since lesion not longer than 6 months, 20 aphasic patients with time since lesion longer than 12 months, 9 patients with diffuse neurological disorders without aphasic symptoms and 15 healthy controls. In all tasks, that were realised within the two-stimulus paradigm, ERP's were characterized by a late positive complex following stimulus onset and a slow negative shift during 2 seconds preceding the S2. The scalp distribution of this slow negative shift varied between experiments and groups, with more frontal and lateralized distribution in the non-acute aphasic group compared to all other groups. When the location of the negative maximum in the scalp distribution of this parameter was established for each subject and task, the non-acute aphasics showed a more pronounced interindividual variability and more lateralized location of maxima than did the controls and the acute group. This was especially prominent in the tasks involving word recognition and phonological encoding. Results suggest that ERPs can be used to examine different aspects of language production in aphasics and healthy subjects. The large interindividual variability in the non-acute group compared to the acute group is interpreted as a result of cortical reorganization and shows that multiple reorganizational processes are possible. From a methodological viewpoint, the interindividual variability may be a useful measure differentiating between groups when averages across groups provide an inadequate impression.

Research was supported by the Deutsche Forschungsgemeinschaft.

## LS/6

EARLY HABILITATION OF DEVELOPMENTAL SPEECH DISORDER IN BRAIN LESIONED INFANTS. *V. Gec[1], J. Ivanus[2]. [1] Center for Hearing and Speech Disabeld, Vinarska 6. 62000 Maribor, Slovenia. [2] University Center for Multidisciplinary Studies, Kneza Viseslava 1. Belgrade, Yugoslavia*

Early dectection of brain lesions in infants may enable to perform habilitation procedures in order to prevent serious and often irreperable speech disorder. Longterm observations indicate the importance of neuroanatomical and neurophysiological escorting. In 30 infants, both sexes up to their second years of age, suffering from perinatal brain lesions were tested for psychomotor abilities and vocalisatin along with performing logopedic treatment. With respect to the pathology of lesion this group was matched to normal speech development and critical vocalization was pointed out. A delay in most syllables was observed. The conclusion of this study is, that propriate habilitation procedures in brain lesioned infants have to be applied from very early neonate days up to maximal success.

## LS/7

THE PROCESSING OF VERBAL AND NONVERBAL EMOTIONAL INFORMATION IN PATIENTS WITH UNILATERAL BRAIN DAMAGE. *A. Geigenberger, M. Schwarz & W. Ziegler. EKN Clinical Neuropsychology Research Group, Department of Neuropsychology, City Hospital Bogenhausen, Dachauer Str.164, D-80992 München*

The issue of emotional information processing in brain-damaged patients is still controversial. Concerning the *processing of nonverbal emotional information*, i.e. prosody and facial expression, a majority of previous studies describes evidence for right hemisphere dominance (Ross, 1993). Other results imply a bilateral processing of nonverbal emotional information, both hemispheres having equivalent or complementary functions (Cancelliere & Kertesz, 1990).
Studies investigating the *processing of verbal emotional information* present divergent results. Some suggest a bilateral localization of emotional processing in general (e.g Gainotti 1993). Others support the view that the right hemisphere plays a predominant role in processing emotion, both verbal and nonverbal (e.g. Borod et al., 1992). The present study was designed to test the right-hemisphere-hypothesis for the processing of verbal and nonverbal emotional information.
**Methods:** The processing *of nonverbal emotional information* was examined by presenting (1) an emotionally ambiguous sentence prosodically realized in six different emotions (Tischer, 1993), and (2) photographs from the *Pictures of Facial Affect* representing five emotional categories posed by male and female actors (Ekman&Friesen, 1976). The stimuli had to be scored on three emotion dimensions (valence, potency, activation) and matched to emotion categories. The processing of *verbal emotional information* was examined using emotional adjectives, metaphors/idioms, sentences with associative primes, and texts containing distractors.
**Sample:** 20 patients with right cerebrovascular lesions (RBD; 16 men, four women); 20 patients with left cerebrovascular lesions (LBD; 13 men, seven women); 34 neurologically healthy subjects.
**Results:** In the processing of *nonverbal emotional information* the RBD patients performed significantly worse than the LBD patients in both the prosodic and the facial expression tasks. As to the processing of *verbal emotional information*, the LBD patients performed significantly worse than the RBD patients. The errors made by the RBD patients could be attributed to a textual processing deficit, in so far as they were unable to draw the required inferences.

### References

Borod, J.C., Andelman, F., Obler, L.K., Tweedy, J.R., Welkowitz, J. (1992), Right hemisphere specialization for the identification of emotional words and sentences. Evidence from stroke patients. Neuropsychologia 3 (9), 827–844
Cancelliere, A.E.B., Kertesz, A. (1990), Lesion localization in acquired deficits of emotional expression and comprehension. Brain and Cognition 13, 171–181
Ekman, P., Friesen W. (1979), Pictures of facial affect. University of California Medical Center, San Francisco
Gainotti, G., Caltagirone, C., Zoccolotti, P. (1993), Left/right and cortical/subcortical dichotomies in the neuropsychological study of human emotions. Cognition and Emotion 7 (1), 71–93
Ross, E.D.(1993), Nonverbal aspects of language. Behavioural Neurology 11(1), 9–23
Tischer, B. (1993), Die vokale Kommunikation von Gefühlen. Weinheim: Psychologie Verlags Union

## LS/8

TWO STAGES IN PARSING: EARLY AUTOMATIC AND LATE CONTROLLED PROCESSES. *A. Hahne & A.D. Friederici. Max-Planck-Institute of Cognitive Neuroscience, Inselstrasse 22–26, D-04103 Leipzig, Germany*

Two ERP-components have been identified to be correlated with syntactic word category processing: an early anterior negativity and a late posterior positivity (Friederici, Pfeifer, & Hahne 1993; Friederici, Hahne, & Mecklinger 1996). These two components are thought to reflect two phases of syntactic processing during sentence comprehension. The early negativity is interpreted as reflecting a first-pass parse defined as the assignment of the initial phrase structure. The late positivity, by contrast, seems to be related to processes of structural reanalysis and repair which may become necessary when the syntactic structure initially build cannot be successfully mapped onto the semantic information and argument information provided by the lexical elements.
We assume that the early negativity reflects a highly automatic process. The present study further examines this automaticity hypothesis by systematically varying the proportion of syntactically incorrect sentences in a given experiment. If this component reflects an automatic process, it should be relatively independent of the participant's concious expectancies and strategic behavior which could be induced by the proportion variation.
In our experiment participants listened to auditorily presented sentences. In one experimental condition 20% of the sentences contained a syntactic word category violation while in another experimental condition 80% of the sentences contained a syntactic word category violation. To the extent that the early negativity is highly automatic and independent of conscious expectancies and strategies, this early ERP-component should be roughly equivalent across proportion conditions. By contrast, to the extent that the late positivity reflects conciously controlled processing, it should vary across the two proportion conditions.
Results showed that the early negativity was present and equally pronounced under both proportion conditions, while the late positivity was observed only for the 20% violation condition, but not for the condition in which participants listened to 80% syntactic word category violations.
This pattern clearly supports the assumption that the early structure building process as reflected in the anterior negativity is highly automatic. The fact that the late positivity could not be observed when proportion of incorrect sentences was high suggests that this process of syntactic reanalysis is much more controlled.

## LS/9

INTRAOPERATIVE LANGUAGE MAPPING: NAMING LATENCY DATA. *J. Ilmberger (1), A. Werani (2), W. Eisner (3), U.D. Schmid (4), H.-J. Reulen (3). (1) Dept. of Physical Medicine, (3) Dept. of Neurosurgery, Klinikum Großhadern, LMU München, Marchioninistr. 15, 81366 München, Germany. (2) Inst. of Phonetics and Speech Communication, Dept. of Psycholinguistics, LMU München, Oettingenstr. 67, 80538 München, Germany. (4) Neurosurgical Unit, Klinik im Park, Seestr. 220, 8027 Zürich, Switzerland*

Introduction
Direct cortical stimulation is used for intraoperative mapping of sensori-motor and language functions. Electrical stimulation may have two effects on behavior: excitatory or inhibitory. Inhibitory effects are observed only in the awake patient as motor or verbal behavior must be self-initiated before it can be inhibited. For motor behavior, it is well known that inhibition may be graded; that is, movement is not stopped completely but only slowed down in ranging degrees. These different forms of inhibition are called negative motor responses. The same negative phenomena have been observed during language testing in intraoperative mapping, e.g. naming may be delayed. However, these effects have never been studied systematically. We report on 2 patients in which naming latencies were recorded during intraoperative mapping of language functions.

## Method

In 2 patients undergoing surgery for removal of space-occupying lesions intraoperative mapping of language functions was performed. Preoperatively, latencies for naming objects were measured. The same set of items was then presented intaoperatively while different cortical sites were electrically stimulated. Disturbances of speech and naming were registered and response latencies were recorded. Intraoperative latencies were substracted from preoperative latencies for individual items; outliers in the difference values were identified by means of confidence intervals resulting from density estimation procedures.

## Results

In the first patient 27 sites were stimulated. At 4 sites electrical stimulation caused a speech arrest, at 3 sites an aphasic (naming) arrest was evoked. At 4 sites prolonged naming latencies were observed, 2 of them lay in near vicinity to sites where arrests of naming occured. In the second patient 28 sites were stimulated. At 2 sites speech arrest was observed, at 7 sites aphasic arrests occured. Prolonged naming latencies were observed at 2 sites, 1 of those sites lay close to a cluster of sites where naming arrests were observed.

## Conclusion

Naming, although normally performed fast and without effort, is a complex process involving several stages. These stages include recognition of an object, activation of semantic and phonological units and articulatory output. Each of these stages may be disturbed seperately under experimental conditions leading to delays in the overt response. It is at present completely unclear at which stage information processing is disturbed by direct cortical stimulation. Moreover, the fact that prolonged latencies are observed at sites that vary considerably in distance from sites at which naming can be blocked completely makes it very likely that stimulation affects different stages of the naming process. However, we were able to demonstrate that registration of naming latencies in the operating room provides a useful method for graded assessment of the importance of specific cortical sites for language processing.

## LS/10

THE ROLE OF THE PERIAQUEDUCTAL GREY IN LIMBIC AND NEOCORTICAL VOCALIZATION CONTROL. *U. Jürgens[1] and P. Zwirner[2]. [1] German Primate Centre, Kellnerweg 4, 37077 Göttingen, Germany. [2] Department of Phoniatrics and Pedaudiology, University of Göttingen, Robert-Koch-Str. 40, 37075 Göttingen, Germany*

A number of observations indicate that the periaqueductal grey of the midbrain (PAG) plays a crucial role in vocal control. Electrical stimulation of this structure yields vocalization in various species. Single-unit recording studies have revealed vocalization-correlated activity in the PAG. Lesions of the PAG can cause mutism. In a recent study, we found that injections of kynurenic acid, an unspecific glutamate antagonist, into the PAG block limbic vocalizations electrically elicitable from various limbic forebrain structures, such as the anterior cingulate cortex, medial amygdala or dorsal hypothalamus. It was concluded from this that the PAG represents a glutamatergic relay station of the limbic vocalization control pathway. These studies, however, did not answer the question of which role the PAG plays in neocortical vocal control. In the motor cortex of several subhuman primate species, there is an area producing vocal fold movements when electrically stimulated. In humans, vocal fold movements as well as complete vocalizations can be obtained from the corresponding region. Bilateral destruction of this region causes loss of voluntary phonatory control.

In the present study, kynurenic acid was injected into the PAG of the squirrel monkey at sites capable of blocking limbic vocalization elicitable from various limbic brain structures. The effect of these injections was studied on the elicitability of vocal fold movements from the motor cortex. It turned out that neither ipsi- nor bilateral injections affected vocal fold movements. These results point to the existence of two separate vocal fold control pathways at midbrain level: one limbic, responsible for non-verbal emotional vocal utterances, and one neocortical, responsible for the production of learned vocal patterns. The PAG represents a crucial relay station of the limbic but not the neocortical vocal control pathway.

## LS/11

COMPARISON BETWEEN VERBAL AND NON-VERBAL SCALES IN THE ASSESSMENT OF MOOD IN APHASIC STROKE PATIENTS. *C. Marra, A. Azzoni[1], F. Gasparini, G. Gainotti. Servizio di Neuropsicologia – Policlinico Gemelli – 00168 – Roma. [1] Servizio Psichiatrico di Diagnosi e Cura, Ospedale S. Spirito – 00100 – Roma*

Several authors have studied the post-stroke depression (PSD). Robinson and co-workers have consistently claimed that: a) there are two forms of PSD: minor and major, b) the major form of PSD is quite similar to the Endogenous Major Depression, c) the major PSD is more frequent in patients with lesion involving the left frontal lobe or the underlying basal ganglia. The main methodological objections against these studies concern the criteria and the scales used to achieve the diagnosis of PSD. Robinson used the DSM-III criteria which are not necessarily valid in stroke patients. Moreover the scales used in the assessment of depression consisted in an interview. Thus, subjects with severe language impairment couldn't be enrolled in the study. To avoid these problems two new diagnostic scales labelled Visual Analogue Dysphoria Scale (VADS) and Post Stroke Depression Scale (PSDS) have been recently developed. The VADS is a measure of subjective sadness involving a 100 mm vertical line with two pictures at the either side of the line representing an happy pole (0 mm) and a sad pole (100 mm). The PSDS is a scale specifically devised to explore PSD. It is composed of ten section taking separately into account many symptoms and behaviour usually observed both in endogenous and in stroke depressed patients. Previous our works have already shown that a)the major PSD shows a behavioural pattern more similar to the minor PSD than to the Endogenous Major Depression, b) there is no correlation between side and location of the brain lesion and the occurrence of Major PSD. The main objection against these works is that patients deeply aphasic could not undergo the PSDS because of their linguistic difficulties, consequently results could be biased by the exclusion of these left brain damaged patients (LBDP). In the present study we analyze the relationship between verbal (PSDS, Hamilton) and non verbal (VADS) scales in the assessment of PSD in anterior-posterior, left-right brain damaged patients. One hundred and thirty-five right handed patients, suffering from unilateral ischemic stroke between two weeks and six months before inclusion, have been assessed for evaluation of mood disorders. Nine LBDP out of 67 could perform only the VADS but not the PSDS because of their linguistic difficulties. Our results show that: a) all the mood measures (verbal and non verbal) are strongly reciprocally correlated both in right and in LBDP, b) there are no differences in the mean score obtained by the patients cross-analyzed for side (left, right) and location of lesion (anterior, central, posterior) at the PSDS and VADS, c) the distribution (anterior vs posterior) of the brain lesions in the nine subjects which could perform only the VADS is well balanced, d)considering also these nine LBDP in the analysis, the VADS mean scores decrease in every group (left anterior, central and posterior). In conclusion the results confirm that there are no differences in the assessment of depression in right and LBDP using verbal and non verbal scales. The analysis of deeply aphasic patients at the VADS shows that these patients are generally more depressed than the other LBDP, however the distribution of the brain lesions in these subjects is not rectricted to the anterior lobe.

**LS/12**

THE TIMING OF COVERT ARTICULATION IN MOTOR SPEE-CH DISORDERS. *W. Ziegler. EKN – Clinical Neuropsychology Research Group, Department of Neuropsychology, City Hospital Bogenhausen, Dachauer Straße 164, D-80992 München*

One of the paradigms used in studies of motor representations draws on comparisons of imagined movements with actually executed actions. The motor actions typically examined are walking, writing, playing tennis etc. In most studies comparable movement times were measured for actual and mental conditions. It is generally inferred that motor imagery and motor execution share a common representational system and are based on overlapping neuronal structures (Jeannerod, 1994). Brain imaging and lesion studies have provided evidence for an implication of parietal, prefrontal lateral and medial, and cerebellar areas in the control of mentally simulated movements.

"Inner speech" has obtained a prominent role in the modelling of spoken language production. It is considered to represent a stage where phonological encoding is completed while the actual execution of motor programs is blocked (Levelt, 1989). Further, subvocal rehearsal plays a major role in current models of working memory. Investigations of covert speech in normals revealed that overt and covert speaking are governed by common structural relations. Studies of speech disordered patients, e.g. anarthrics or stutterers, tried to establish a relationship between the underlying pathology and the nature of inner speech representations. Activation studies yielded conflicting results as to the role of specific brain regions (e.g. Ryding et al., 1996). Regions of interest are the SMA and the inferior frontal cortex of the dominant hemisphere and the cerebellum.

This study used a counting task to examine temporal aspects of covert articulation in patients with neurogenic speech disorders, i.e. after lesions to the cerebellum (N = 2), brainstem (N = 1), Broca's area (N = 2), left prefrontal cortex (N = 1), and left SMA (N = 1). Articulatory requirements were varied in a three factor orthogonal design including (1) overt vs. covert speech, (2) "clear" vs. "rapid" articulation, (3) counting "from 1–13" vs. "from 21 to 25". Factors (2) and (3) were included to control for rate- and accent-related invariance.

All patients except the tachylalic were slowed in both overt and covert speech. In the normals, the time relations imposed by factors (2) and (3) were preserved in covert speech. Only one apraxic patient violated this invariance principle. Most patients experienced similar problems in covert as compared to overt speech. It is concluded that covert articulation involves cerebellar as well as left frontomedial and frontolateral regions. Attentional influences in the covert condition are suggested to have reduced the disinhibition problem in the tachylalic subject and to have increased the programming deficit in the apraxic speakers. The results have implications for the role of mental training techniques in the treatment of disordered speech.

**References:**
Jeannerod, M. (1994). The representing brain: neural correlates of motor intention and imagery. *Behavioral and Brain Sciences*, 17, 187–202
Levelt, W.J.M. (1989). Speaking. From Intention to Articulation. Cambridge: MIT-Press
Ryding, E., Brådvik, B., & Ingvar, D.H. (1996). Silent speech activates prefrontal cortical regions asymmetrically, as well as speech-related areas in the dominant hemisphere. *Brain and Language*, 52, 435–451

# Movement

## MO/1

CONTINUOUS b-AMYLOID$_{(1-42)}$ INFUSION INDUCES BEHAVIORAL, BIOCHEMICAL AND NEUROANATOMICAL CHANGES IN FREELY MOVING RATS: NEUROPROTECTION BY ZM-6, A PENTAPEPTIDE ANTAGONIST. *Ábrahám, I.[2,4], Timmerman, W.[5], Rensink, A.A.M.[2], Sasvári, M.[1], Varga, J.[3], Jost, K.[3], Zarándi, M.[3], Penke, B.[3], Nyakas, C.[1,2], Harkany, T.[1,2] and Luiten, P.G.M.[2]. [1] CRD Haynal University of Health Sciences, Budapest, Hungary, Depts. of [2] Animal Physiology and [5] Pharmacology, University of Groningen, The Netherlands, [3] Dept. of Medical Chemistry, Szent-Györgyi Albert Medical University, Szeged, Hungary and [4] Institute of Experimental Medicine, Budapest, Hungary*

β-amyloid (bAP) neurotoxicity, *in vivo* properties of potentially amyloid-like peptide fragments and their probable contribution to cell death in brain areas subserving learning and memory have recently become one of the major issues in Alzheimer's disease (AD) research. Previously we demonstrated that bAP fragments exhibit *in vivo* cholinotoxicity in the magnocellular nucleus basalis (MBN) of rats. In the present experiments we investigated a 120-min microdialytic bAP infusion-induced changes in animal behavior throughout a 14-day survival period. Alterations in the releases of excitatory amino acids (EAA) during bAP perfusion were also determined. To evaluate the neurotoxic potential of bAP, acetylcholinesterase (AChE) histochemistry was used. Effects of a pentapeptide amyloid antagonist (ZM-6) which may prevent the cell-surface binding or aggregation of bAP were also studied.

A microdialysis probe was implanted into the right MBN region of male Wistar rats and bAP$_{(1-42)}$ was dialysed (0.2 nmol/ml) in the freely moving animals. In neuroprotection studies, an equimolar ZM-6 solution was dialysed for 60 min followed by a 60-min bAP$_{(1-42)}$ perfusion. EAA levels were determined by HPLC analysis. In parallel with bAP perfusion home cage-like activities were determined. Furthermore, open-field activity was recorded after 24 hours, while passive shock-avoidance learning was tested 14 days post-surgery. Thioflavine-S histochemistry was employed to visualize the extent of bAP$_{(1-42)}$ deposition and the loss of AChE-positive fibers in the somatosensory cortex was recorded by means of quantitative histochemistry.

Prolonged bAP administration resulted in impairments of both spontaneous behaviors and learning and memory processes. Pretreatment with ZM-6 beneficially influenced these abnormalities. bAP infusion increased the extracellular concentration of aspartate, glutamate and taurine. Interestingly, the time point of this enhanced EAA release matched the activation in home cage-like behavior. Comparison of the loss of cortical AChE-positive fibers revealed a strong correlation between the tertiary structure of bAP fragments and the extent of bAP depositions. ZM-6 administration attenuated the enhanced EAA release and prevented the loss of cholinergic fibers in the somatosensory cortex.

In summary, the reason of bAP$_{(1-42)}$-elicited marked behavioral disturbances throughout the survival period might be the extensive neurodegeneration. Early phases of *in vivo* bAP neurotoxicity involve an enhanced release of EAA directly leading to cell death. Furthermore, conformational amyloid antagonists effectively prevent the initiation of the neurotoxic cascade and might be potent tools in designing specific anti-amyloid drugs.

## MO/2

MAY BASAL GANGLIA BE INVOLVED IN CONNECTING SOMATOMOTOR RESPONSES WITH CARDIORESPIRATORY CHANGES? *L. Ángyán. Institute of Physiology, Medical University of Pécs, Szigeti út 12, 7643 Pécs, Hungary*

It is generally accepted that somatomotor activities are associated with appropriate changes in the cardiorespiratory functions to fulfill the metabolic demands of those motor activities. However, the neural mechanism of this connection is not yet clear, though both peripheral and central mechanisms have been suggested. The basal ganglia circuitry is known to play important role in motor control. Considering the widespread connections of this group of nuclei, the present study questioned whether the basal ganglia may be involved in connecting somatomotor responses with appropriate cardiorespiratory changes. The somatomotor and cardiorespiratory responses to electrical stimulation within the basal ganglia circuitry were studied in freely moving cats with chronically implanted electrodes. Electrodes were placed in the caudate nucleus (CD), putamen (Put), globus pallidus (GP), subthalamic nucleus (Sub) and substantia nigra (SN). We describe locus-dependent increases in the arterial blood pressure (BP), heart rate (HR) and respiratory rate (RR) that are obtained from freely moving cats. Generally, slight changes in cardiorespiratory responses were produced by stimulation of CD and Put, while significant increases in BP, HR and RR were obtained from the other loci. The high amplitude increase in BP, HR and RR failed to occur during stimulations repeated under the blockade of adrenergic a$_1$-receptors by phentolamine, or under chloralose anesthesia, as well as after local injection of a neurotoxin, kainic acid. It is concluded that basal ganglia are able to modify the cardiorespiratory functions simultaneously with the somatomotor behaviour.

## MO/3

EVALUATION OF FAST fMRI MEASUREMENTS IN MOTOR-CORTEX: COMPARING A NEW CLUSTER ANALYSIS WITH z-MAPPING. *A. Baune[1,2], M. Erb[1], F. T. Sommer[2], D. Wildgruber[1], U. Klose[1], W. Grodd[1]. [1] Section Exp. MR of the CNS, Dept. of Neuroradiology, Hoppe-Seyler Str. 3, D-72076 Tübingen. [2] Dept. of Neural Information Processing, University of Ulm, D-89069 Ulm*

**Introduction:** Fast functional magnetic resonance (fMRI) measurement with single shot echo-planar imaging (EPI) detected sequential activation of SMA and primary motor cortex (M1) during voluntary finger movement (Erb et al). We propose a new postprocessing method using dynamical clustering analysis examining the entire time courses of the pixel and compare it with standard z-mapping. **Methods:** fMRI was performed on a 1.5 T scanner using a single slice EPI sequence (TE = 43 ms, a = 40°, FOV = 192 mm, 64 × 64 matrix, slice thickness 4 mm) with a repetition time (TR) of 107 ms. In each experiment 1024 subsequent images were acquired for a time period of 110 s. Four healthy, right handed volunteers were instructed to press a button repeatedly with the right index finger during presence of a light signal. Four times the light was switched on for 5 s after a 20 s pause interval and five of these measurements were performed.

Postprocessing of the data averaged over the five measurements: The standard method used the z-map to classify activated pixels which in addition must belong to an area of at least 5 adjacent pixels with mean signal difference above 0.5 standard deviation (activation kernel). Two regions of interest (ROI) were determined within the acquired axial slice corresponding to SMA and M1. The mean signal in each ROI was crosscorrelated with a action step function to obtain the time delay.

Alternatively, a dynamical unsupervised clustering analysis was applied. The number of clustering centers was adjusted by two thresholds (cluster generation and fusion distance). The activation of a cluster center is given by the crosscorrelation maximum with the action step function. Activated clusters contribute to the activation of a ROI if more than half of the cluster pixels belong to the ROI. The activation delay of a ROI was determined by crosscorrelation of the averaged time course of all contributing cluster centers with the action step function.

**Results:** All high activated clusters contribute to SMA or M1 with three exceptions: 1) we observed single activated pixels in ipsilateral M1, 2) in visual cortex and 3) in the sagittal sinus. In all four volunteers we found activated clusters belonging to SMA and M1 as well. This suggests the presence of early components in M1 synchronious with SMA. We measured time shifts between M1 and SMA activation: 0.2 s, 0.5 s, 0.2 s and 0.2 s which were in the range of the reaction time. The shift values obtained by standard postprocessing were larger, in one case far from being plausible: 0.4 s, 1.7, s 0.3 s and 0.4 s.

**Conclusions:** The structure of activated clusters provides new insights by discriminating different time courses within ROIs and by detecting activation outside. By analyzing the entire time courses the proposed dynamical cluster analysis method confirms the results obtained by standard postprocessing: For the considered experiment the predefined ROIs really correspond to the most activated brain regions (which cannot be decided by z-mapping). These results are encouraging to apply the cluster method for experimental paradigms where ROIs cannot be defined in advance. Quantitatively, the new method yields shorter (and also more consistent) time shifts between SMA and M1 actiavation.

## MO/4

GROOMING SYNTAX DISRUPTED IN D1A MUTANT MICE. *H.C. Cromwell*, K.C. Berridge†, C.A. Crawford*, J. Drago‡ and M.S. Levine*. * Mental Retardation Research Center, University of California at Los Angeles, Los Angeles, CA. 90025; † Department of Psychology, University of Michigan, Ann Arbor MI. 48109; ‡ Laboratory of Mammalian Genes and Development, National Institute of Child Health and Development, NIH, Bethesda MD 20892*

The role of dopamine in movement processing is not well understood. A complicating factor in our understanding is that the dopamine system has a multifarious receptor family. The dopamine system has five known receptor subtypes, and the different subtypes have been hypothesized to make contributions to different types of behaviors. Activation of the D1 receptor subtype has been intimately linked with certain behaviors, and one well known example is D1 receptor activation leading to grooming activity in rodents. The goal of the present study was to examine the ordered grooming chain in mutant mice lacking the D1A receptor. An animal model to examine dopamine function has recently been devised by using gene targeting with homologous recombination to produce mutant mice which lack a functional D1A receptor. Rodent grooming is composed of both flexible and inflexible periods of activity. The shorter duration and less frequent inflexible grooming bouts include a series of four action phases tightly ordered in a reliable pattern. This reliable pattern has been termed the grooming chain, and its syntax is sensitive to both excitotoxin lesions of the striatum and midbrain dopamine neuron destruction. The integrity of the grooming chain was determined by examining the completion percentage of the mutant population (−/−) and comparing it to the wild-type control's (+/+) completion percentage. Before using the mutant mice for behavioral studies, they were genotyped using southern analysis. After identification, mice were given a new number designation and filmed by a person blind to the genetic makeup of the animal. All grooming bouts were filmed in a ten minute period. After the initial 5 minutes, a light water spray was used to elicit more grooming. Grooming tapes were scored by slow motion video analysis by another person blind to the mice genotype. Results showed that mutant mice lacking the D1A receptor had a significant impairment in grooming chain completion compared to controls ($40.8 \pm 3.12\%$ vs. $54.2 \pm 6.5\%$ in mutants and controls, respectively, Mann Whitney U Test, $P < 0.05$). In several instances, mice showed interesting syntax distortions which reflect a degraded kind of completion because all four phases occur yet they do so out of the natural order. The results of an impaired action sequencing in these mice support the idea that D1A receptor activation is important for appropriate movement output and that this receptor plays a specific role in the linking together of natural actions. The results have implications for our understanding of certain neural disorders with a deficit in action sequencing as a symptom such as Parkinson's disease and may speed the process of treating such functional problems.

## MO/5

WHOSE HAND IS THIS? *Elena DAPRATI, CNRS UPR 9075, Nicolas FRANCK CNRS UPR 9075, Nicolas GEORGIEFF CNRS UPR 9075, Marc JEANNEROD CNRS UPR 9075. Institut des Sciences Cognitives CNRS UPR 9075, 8 Av Rockefeller, 69373 Lyon Cedex 08, France*

We investigated the mechanisms leading to conscious awareness in self-generated movements by analysing the performance of 12 healthy subjects in a recognition task.
Subjects were required to execute simple finger and wrist movements with their right hand. The moving hand was not directly visible, but its image was captured by a camera, and shown to the subjects in real time on a TV screen. This artifice allowed the experimenter to occasionally replace the image of the subject's hand with that of his own hand. Hands wore an identical glove, in order to minimise differences due to physical features. The task for the subjects was to perform a movement, and once it was over, to decide whether the hand they had seen on the TV screen was their own or not. Answer was given by a YES or NO verbal statement. In separate blocks, two sets of four movements each were used: Finger Set (starting position: grasp; 1. extend thumb, 2. extend index finger, 3. extend index and middle fingers; 4. open the hand wide); Wrist Set (starting position: hand around a lever; 1. push forward, 2. push backward, 3. push leftward, 4. push rightward). In each set, three different conditions were tested: Subject (the subject was shown his/her own hand); Experimenter Same (the subject was shown the experimenter's hand performing exactly the same movement); Experimenter Different (the subject was shown the experimenter's hand performing a different movement).
The number of incorrect responses (error rate) was recorded for each subject. Results were submitted to an ANOVA, whose within-subjects' factors were: Movement Set (Finger Set, Wrist Set) and Hand Presented (Subject, Experimenter Same, Experimenter Different). Subjects seldom failed to detect the alien hand in condition Experimenter Same. That is, when the image of the subjects' hand was substituted by that of the experimenter performing the same movement, subjects occasionally misattributed it to themselves. This gave rise to 4.9 errors in average, with respect to 0.1 errors and 0 errors in the condition Subject and Experimenter Different, respectively ($F(2,22) =$, $P < 0.00001$). Experiments extending this paradigm to pathological populations (e.g., schizophrenics) are under way.

## MO/6

SEQUENTIAL ACTIVATION OF SMA AND M1 DURING SELF-PACED FINGER MOVEMENT DETECTED BY fMRI: ANALYSIS WITH CORRELATION AND CLUSTER METHODS. *M. Erb[1], D. Wildgruber[1], A. Baune[1,2], F. T. Sommer[2], U. Klose[1], W. Grodd[1]. [1] Section Exp. MR of the CNS, Dept. of Neuroradiology, Hoppe-Seyler Str. 3, D-72076 Tübingen. [2] Dept. of Neural Information Processing, University of Ulm, D-89069 Ulm*

**Introduction:** From cortical DC potentials a characteristic spatio-temporal pattern of slow negative voltage shift in the mesial, fronto-central cortex is known, which precedes any kind of voluntary movements. This shift starts about 0.5–1 s prior the actual motor potential and is attributed to electrical activity of the supplementary motor area (SMA). As functional magnetic resonance (fMRI) is well able to depict activated cortical structures with high spatial but limited temporal resolution, we have applied repeated single shot echo-planar-imaging (EPI) to detect sequential activation of SMA and primary motor cortex (M1) during voluntary finger movement. In contrast to our earlier experiments, we have analyzed signals from single self-paced events. Evaluation of activated pixels was performed both with correlation analysis and with a new developed cluster method.

**Methods:** fMRI was performed on a 1.5 T scanner (Siemens Vision) using a single slice EPI sequence with a repetition time of 107 ms. In each experiment five trials with 1024 subsequent images were acquired for a time period of 110 s. Seven healthy, right handed volunteers were examined and instructed to press a button with the right index finger in a self-paced speed approximately every 10–15 seconds.

As conventional postprocessing we used cross-correlation after averaging the image series according to the time of actual finger movement. Pixels were declared as activated if they belonged to an area of at least 9 adjacent pixels with a correlation coefficient (cc) > 0.3. Regions of interest (ROIs) were determined within the acquired axial slice corresponding to the SMA and M1. The time course of mean signal intensity within these ROIs were calculated and cross-correlated with a step function. The time shift between the fMRI signal in the different ROIs and the finger movement was defined as the time shift with the maximal cc. The timeshift for the different areas were tested with a t-test for paired samples.

The new developed postprocessing method classifies activated pixels, by a dynamic unsupervised clustering algorithm. The number of resulting clusters was determined by two thresholds (generating distance and fusion distance). Cluster centers showing high correlation coefficients with the step function were used to calculate the time course of the mean signal intensity within each ROI.

**Results:** The mean time delay of SMA activation compared to the onset of finger movement was 3.0 s (SE = 0.6s) and in the M1 3.8 s (SE = 0.6s), the differences between both regions were statistically significant (p < 0.0001), the mean time difference (Dt) was 0.75 s (SE = 0.1 s).

**Conclusions:** Our results demonstrate that it is possible to achieve a sufficient temporal resolution in fMRI to detect the sequential activation of functionally connected cortical areas of the motor system. The time shift of 0.75 s between the hemodynamic response of the SMA and M1 measured with fMRI is in very good agreement with the electrophysiological data, supporting the hypothesis that vascular response parallels the sequence of electrical activation in the cortex.

## MO/7

PEAK FORCES SPECIFY THE TIMING OF ACTIONS: SYNCHRONIZING ISOMETRIC FORCE PULSES. *J. Gehrke, G. Aschersleben, & W. Prinz. Max-Planck-Institut fr psychologische Forschung, Leopoldstr. 24, D-80802 Mnchen, Germany*

In synchronization tasks subjects are instructed to synchronize a motor response (e.g., taps) with a regular sequence of stimulus events (e.g., clicks). The relationships between the onsets of the pacing signals and the onsets of the response signals are then measured. Usually, a so-called negative asynchrony is observed, that is, the motor event precedes the sensory event by 30–50 ms. An account explaining the negative asynchrony is based on two assumptions: First, synchrony is established at a central level, that is, central representations of taps and clicks are brought to coincidence and not their corresponding external events in the world. Second, the motor response is represented by its (sensory) effects. Afferent rather than efferent movement codes are superimposed on the afferent codes representing the pacing signal. The movement-related afferent codes result, for example, from tactile/kinesthetic feedback of the tap. Thus, the negative asynchrony occurs as a result of the differences in processing times between the two codes (Aschersleben, & Prinz, 1995).

Recent experiments investigating the tap-induced stimulation in more detail showed that the size of the negative asynchrony is dependent on the intensity of the tactile/kinesthetic stimulation. Increased peak forces and movement velocities go along with reduced asynchronies as compared to conditions with weaker forces (Gehrke, Aschersleben, & Prinz, 1997). The explanation given for this effect is that the time to generate a central representation is reduced under conditions with increased stimulation of the finger. To test whether this explanation can be generalized to different synchronization tasks we conducted experiments using isometric force pulses, that is, subjects were instructed to synchronize isometric taps instead of isotonic taps (which include a lift of the finger). The isometric tapping conditions revealed results similar to the ones observed with isotonic taps. Under conditions with increased isometric force pulses the asynchrony is reduced, thus, indicating that the timing of the tap

is dependent on the intensity of the stimulation. In addition, different to the results obtained by Freund and Büdingen (1978) time to peak force was dependent on peak force. As comparable results were observed in both the isometric and the isotonic tapping tasks we conclude that the relation between the intensity of the action effect and timing of the motor act is a general one.

Aschersleben, G. & Prinz, W. (1995). Synchronizing actions with events: The role of sensory information. Perception & Psychophysics, 57, 305–317

Gehrke, J., Aschersleben, G., & Prinz, W. (1997). Processing of afferent feedback and the timing of actions: Evidence for a sensory accumulator model of synchronization. Ms submitted for publication

Freund, H.-J., & Büdingen, H.J. (1978). The relationship between speed and amplitude of the fastest voluntary contractions of human arm muscles. Experimental Brain Research, 31, 1–12

## MO/8

COMPENSATORY LEG MUSCLE REACTIONS FOLLOWING STANCE PERTURBATIONS: INFLUENCE OF CORTICO-SPINAL INPUT. *M.E. Keck[1,2], M. Schubert[3], M. Pijnappels[4], A. Curt[1], G. Colombo[1], V. Dietz[1]. [1] Swiss Paraplegic Centre, University Hospital Balgrist, University of Zurich, 8008 Zurich, Switzerland. [2] Max-Planck-Institute of Psychiatry, Clinical Institute, 80804 Munich, Germany. [3] Dept. of Clinical Neurology and Neurophysiology, University of Freiburg, 79106 Freiburg, Germany. [4] Dept. of Med. Physics & Biophysics, University of Nijmegen, 6525 EZ Nijmegen, The Netherlands*

During the performance of functional movements rapid, automatic muscle activation is required to compensate for unexpected external disturbances. Using the transcranial magnetic stimulation (TMS) technique it has been shown that long-loop reflex mechanisms are involved in the compensatory reactions of hand and forearm muscles. Until now the same technique has not been applied yet to study corresponding mechanisms in lower leg muscles. Therefore we studied the interaction of supraspinal motor centers with the generation of compensatory leg muscle responses following stance perturbations. TMS (MagPro, Dantec) just below motor threshold was applied randomly at 19 different time intervals before and during the onset of stance perturbation and for comparison during an equivalent voluntary foot dorsiflexion task. Electromyographic (EMG) activity from the tibialis anterior (TA) muscle and corresponding ankle joint movements were recorded from both legs. To ensure stability of the stimulation procedure control recordings were performed from the abductor digiti minimi muscle under the same conditions. Forward directed displacements were induced by randomly timed ramp impulses of constant acceleration upon a custom-built moveable platform. For comparison, leg muscle EMG was recorded during dynamic isometric foot dorsiflexion during stance while leaning against a support. The stance perturbations were followed by a compensatory response (CR) in the TA with a mean onset of 81 msec. Only in the voluntary task an early facilitation (pre-movement facilitation) of the evoked motor responses (EMR) in the TA following TMS was found which started 40 msec *before* the onset of the EMG (p < 0.05) and lasted over the dynamic part of the TA EMG activity. In the stance perturbation task the TA EMR was facilitated to a *lesser* degree (p < 0.005) and paralleled the increasing slope of the CR. Following the dynamic phase of the TA – CR (about 40 msec after onset) there was no significant difference in the amount of EMR facilitation between stance perturbation and the voluntary task. We assume that during this plateau phase the facilitation had reached a steady-state in both tasks. Opposed to these results a stronger facilitation of the EMR was described for hand muscles under corresponding conditions of automatic compensation for muscle stretch, strongly suggesting a transcortical reflex loop. Thus a spinal generation of the TA – CR following stance perturbation can be assumed when it is necessary that leg muscle activation occurs quickly and in an appropriate way to ensure postural stability. Taken together, these results point towards a differential neuronal control of leg and hand muscles.

Supported by Swiss National Foundation, International Research Institute for Paraplegia and STIR (University of Nijmegen)

## MO/9

**fMRI OF CERBRAL ACTIVATION DURING IMAGINED HAND MOVEMENTS:** *M. Lotze[1], P. Montoya[2], U. Klose[1], B. Kardatzki[1], N. Birbaumer[2], W. Grodd[1]: [1] Section Exp. MR of the CNS, Dept. of Neuroradiology, Hoppe-Seyler Str.3, D-72076 Tübingen; [2] Institute of Medical Psychology, Gartenstrasse 29, D-72074 Tuebingen*

**Object of the study:** The involvement of premotor cortex and primary motor areas (M1) and the cerebellum in the imagination of voluntary movement is still uncertain. For frontal areas like the supplementary motor area (SMA), the prefrontal cortex and the gyrus cinguli an increase in activation is found during imagination of hand movements compared to executed movements in Positron-Emission-Tomography (1), while for the cerebellum no consistent data are reproted. We therefore tried to investigate the cerebral perfusion during voluntary and imagined movement of both hand with functional magnetic resonance imaging (fMRI) of the brain to assess changes in the frontal lobe, the somatosensory cortex (S1) as well as in the cerebellum.

**Material and Method:** fMRI-data of the whole brain were acquired with a commercial 1.5 Tesla tomograph (Siemens Vision) using a multislice echo planar imaging sequence with 27 axial slices (4 mm slice thickness, 1 mm gap, 128*96 matrix, 260*162 mm FOV, TE 66 ms, TR 1,8 ms, acquisition time 4 sec). Sequential real and imaginary hand movements were performed by 10 right-handed volunteers (5 male and 5 female, 19–40 years). Executed hand movements (EM) consisted of pressing a softball with the right or left hand. Imagined movements (IM) of both hands were trained prior to the fMRI examination under EMG control and with a personal assessment score of intensity of imagination. Statistical evaluation of the fMRI data was performed by calculation of z-values, where pixels with a z-value of ° 5.6 were declared as activated if localized in anatomically defined regions of interest (ROI) [M1, S1, SMA, cerebellar hemispheres and vermis].

**Results:** During IM activation in the contralateral M1 and S1 is reduced to appr. 30% compared to EM, while the activation for SMA remains the same. In the cerebellum in all cases activation is found in the ipsilateral anterior and posterior lobe. For the anterior lobe the activation during EM revealed significant differences between left and right hand (65% higher activation for left hand movement). For IM this activation in the ipsilateral anterior lobe was diminished by 70% for the right and 85% left hand. In the posterior lobe a less prominent activation in respect to the anterior lobe (35% for right hand and 20% for left hand) was found during EM. For IM this activation is decreased by 25% for the right hand and by 70% for the left hand.

**Conclusion:** Our results support fMRI observations, which describe lower activation in M1 and S1 and no change in SMA activation during IM (2). In the cerebellum a strong selective decrease of activation in the ipsilateral anterior cerebellar lobe is found during IM in comparison to EM for both hands. This reduction may depend on the lack of afferent information during IM in the anterior lobe. The detected activation in the posterior lobe may indicate that the cerebellum is also involved in planning and control of movement. Interestingly this activation site did not reveal differences between EM and IM for the dominant hand, which may attributed to a persistent efferent activity in both states.

**References:**
1. Stephan K.M. et al.: 1995, Journ. of Neurophysiology, 73:373
2. Leonardo M., et al.: 1995, Human Brain Mapping, 3:83

## MO/10

**TRANSCRANIAL MAGNETIC STIMULATION SELECTIVELY IMPAIRS INTERHEMISPHERIC TRANSFER OF VISUO-MOTOR INFORMATION IN NORMAL HUMAN SUBJECTS.** *A. Maravita, C.A. Marzi, C. Miniussi, J. Rothwell[*], J. Sanes[#], L. Bertolasi and G. Zanette. Dept. of Neurological and Visual Sciences, Strada Le Grazie, 37134 Verona, Italy; [*] MRC HMBU Inst. Neurol., Queen Sq., London WC1N 3BG, U.K.; [#] Dept. Neurosci., Div. Biol. & Med., Brown Univ., Providence, RI 02912, USA*

In a simple reaction time (RT) task subjects are typically slower in responding to visual stimuli presented to the hemifield contralateral to the responding hand than to the ipsilateral hemifield. In the latter condition stimulus detection and motor response can be subserved by one and the same hemisphere while in the former both hemispheres are involved. Poffenberger (1915) was the first to hypothesize that an interhemispheric transfer (IT) via the corpus callosum was responsible for the difference between crossed and uncrossed visuo-motor combinations, the so-called CUD. Such an hypothesis has been strengthened by the finding that the CUD shows a multifold increase in subjects with callosal agenesis and, to an even greater extent, in surgically callosotomized patients.

In the present study we sought to test more directly the callosal hypothesis of the CUD by using transcranial magnetic stimulation (TMS) in 10 normal subjects performing manual RTs to LED-generated brief flashes randomly presented to the hemifield contralateral or ipsilateral to the responding hand. TMS (1.5 Tesla; 80% of stimulator output) was applied 50 msec after the flash with a 8-shaped coil positioned in the scalp over the right or left occipital bone 2 cm lateral and 2 cm above the location of the electrode Oz according to the International 10–20 system. All hemifield, hand and side of TMS combinations were randomized and catch trials were introduced in which no TMS but only the noise produced by the coil was administered following visual presentations. In other catch trials only the acoustic and magnetic stimuli were presented but no visual stimuli. The results showed a reliable effect of TMS in retarding (mean:12 - msec) RTs only in the crossed conditions. No such effect was exerted by acoustic stimulation alone and therefore we believe that our results can be explained by an interfering effect of TMS on IT. Such an effect tended to be more marked when the TMS coil was positioned over the hemisphere ipsilateral to the hemifield of stimulus presentation and this suggests an interference at the level of the cortical areas receiving rather than sending the callosal input.

## MO/11

**PSYCHOPHYSICS OF KINAESTHESIA: DIFFERENCE THRESHOLDS FOR THE PERCEPTION OF THE VELOCITY OF VOLUNTARY LIMB MOVEMENTS.** *Müller, St. & Wist, E.R. Institut für Physiologische Psychologie II, Heinrich-Heine-Universität Düsseldorf, Universitätsstraße 1, 40225 Düsseldorf*

How well are differences in the the velocity of limb movements perceived? Limb velocity dependent neurones have been found in the parietal and frontal cortex. Neurographic studies have demonstrated limb velocity dependent modulation of discharge in afferent nerves. Muscle spindels are favourised as responsible receptors. Velocity perception can selectively be disturbed by tendon vibration and does not rely on a simple trade-off between amplitude and movement duration. Quantitative and qualitative differences in the responses of neurones to passively guided and actively executed movements have been found in single cell studies.To date only a few psychophysical studies exist on the perception of differences in the velocity of passive limb movements and none, to our knowledge, concerning voluntary active movements.

We measured difference thresholds for the perceived velocity of voluntary arm movements for target velocities of 10, 20 and 30 cm/s (equivalent to a range of 14–50 deg/s). Somatosensory, vestibular, visual and auditory feedback was excluded. Six healthy human subjects performed 500 extension movements with their dominant right arm in three separate sessions, one target velocity per session. Amplitude and duration of movements were not restricted in order to avoid judgements of amplitude and/or time per se. Subjects were required to produce a constant velocity within each trial which corresponded as exactly as possible to the target velocities. After each trial the subjects made a forced choice response concerning whether the produced movement was faster or slower than the target velocity . The slope of one regression line presented on a monitor represented target velocity in cm/s. A second regression line served as feedback for the actually produced movement velocity. A visual comparison of the two slopes provided feedback on both produced and judged velocity. A monetary bonus system was used which maximized pay-off for the combined accuracy of matched and judged velocites and resulted in a significant improvement in performance. The bonus was presented at the end of each trial together with the visual feedback.

Psychometric functions were calculated on the basis of the 500 measurements obtained for each of the three target velocities and for each subject. Mean 75%-difference thresholds increased linearly as a function of target velocity from 1.28 cm/s (10 cm/s), to 1.73 cm/s (20 cm/s) and 2.24 cm/s (30 cm/s). The corresponding Weber fractions were 0.128, 0.087, and 0.075.

To our knowledge, these data represent the first determination of difference thresholds for the kinaesthetic perception of the velocity of limb movements. This is surprizing because the definition of the term "kinaesthesia" involves the perception of movement. In spite of this fact, most studies have concentrated either on the position sense or on detection thresholds for limb movements.

## MO/12

BIMANUAL COORDINATION IN A PATIENT WITH CALLOSAL DYSPLASIA AND COLPOCEPHALY: FURTHER SUPPORT FOR CALLOSAL MODULATION OF SUBCORTICAL INTERACTIONS. *B. Preilowski[1], P.Schellig[2]. [1] Department of Psychology, Tübingen University, Weissenau Fieldstation, Rasthalde 3, D-88214 Ravensburg, Germany. [2] Neurological Rehabilition Centre Jugendwerk Gailingen, D-78260 Gailingen, Germany*

The results of several tests of bimanual coordination are reported for an 18-year old woman (M.L.) with dysplasia of the corpus callosum. In an MRI only the anterior commissure and a small anterior portion of the callosum could be identified. The rostrum was not visualized and the cingular gyrus appeared poorly developed. Both occipital horns were disproportionately enlarged in comparison with other parts of the lateral ventricles which is suggestive of colpocephaly. Results of a routine neurological examination of M.L. were within normal ranges as were data from intelligence testing. The findings of bimanual coordination tests differed from those of patients with more complete surgical or pathological lesions of the neocommissures and those of patients with callosal agenesis. The main difference was the ability of M.L. to perform simultaneous movements with both arms and hands at different speeds and directions with and without visual control. In contrast with this ability were difficulties in performing mirror image movements at greater speeds.

The present summary concentrates on the results of a test which requires two crank handles to be simultaneously turned in order to move a lightspot on a monitor. Turning with the right hand produces vertical and turning with the left hand horizontal movements. Mirror image as well as parallel movements were required in specific trials. Another variable was the ratio of simultaneous left to right hand input. The interpretation of the results is based on ideas about callosal functions derived from studies with patients with various degrees of surgical or pathological callosal lesions performing similar tests and from animal experiments. These ideas emphasize the role of callosal interhemispheric inhibition, the possibility of regional and state dependent changes in callosal functions (e.g. with practice) and the interaction of callosal and subcortical mechanisms during bilateral coordination (Preilowski, 1990; Preilowski, 1995).

In the case of M.L. commissural transfer appears sufficient for bilateral coordination to reach a fair but not perfect level. Thus, for example, the tendency of mirror image associated movements can be suppressed and – supposedly through the exchange of motor corollary information – coordination becomes independent of visual control. However, while healthy controls very soon are able to pay little attention to one or the other hand which is moved at maximum speed, while concentrating on the other hand to contribute just enough to keep a specific direction, M.L. appeared unable to achieve complete independence of either hand. Her movements always remained very consciously controlled. This was especially apparent in the mirror symmetrical condition during which normal controls achieve maximum speeds supposedly using subcortical possibly even spinal mechanisms with minimal conscious control. M.L. does not profit from such a condition, apparently unable to overcome bi-

hemispheric control because of a lack of sufficient callosal interhemispheric inhibition.

Preilowski, B. (1990). Intermanual transfer, interhemispheric interaction, and handedness in man and monkeys. In C. Trevarthen (Ed.), Brain circuits and functions of the mind. Essays in honor of Roger W. Sperry, (pp. 168–180). Cambridge: Cambridge University Press

Preilowski, B. (1995). Functions of the corpus callosum in interhemispheric interaction: transfer and inhibitory modulation. Society for Neuroscience Abstracts, 21, 1423

## MO/13

EFFECTS OF ATTENTIONAL CUEING ON MANUAL PREHENSION – A KINEMATIC ANALYSIS. *C.L.Pritchard and A.D.Milner. University of St. Andrews, St. Andrews, Fife, Scotland KY16 9JU*

Previous studies have shown that subjects are faster at detecting a target if they are first cued to the target location. This is true even when the cue summons attention covertly, i.e. in the absence of eye movements. In most of these studies subjects respond by pressing a single key as soon as they detect the target. However, several recent studies have investigated the relationship between spatial attention and goal-directed action (Sheliga et al, 1997; Howard and Tipper, 1997). They have shown that stimuli which are to be ignored can influence the trajectories of both saccades and manual responses. This is true both for distractors which are ignored throughout a trial, and for stimuli which act as cues, i.e. stimuli which the subject must first attend to before moving attention to the target. The present study used a modified version of the Posner cueing paradigm to investigate the effects of attentional cueing on subjects manual responses. Targets were opaque cylinders, one placed 14 cm to the right of the subjects sagittal midline, the other 14 cm to the left. Red LED's inside the targets meant that they could be illuminated. Subjects were instructed to reach out, as quickly as possible, and grasp the target which became illuminated. Prior to target illumination subjects were first cued, either to the side of the target (valid cue) or to the opposite side (invalid cue). Cues were green LED's embedded in the work surface. Reaction time and movement time were recorded, as well as various kinematic markers of the reach. These were: peak velocity, time to peak velocity, peak acceleration and percentage of the reach spent decelerating. The results will be discussed in terms of the effects of spatial attention on target detection and response planning and on the kinematics of the reach itself. Data will also be presented on maximum grip aperture (and time to maximum grip aperture). Milner et al (1992) have presented data showing that, in normal subjects, cueing can cause a misperception of a mid-bisected line, whereby the cued half looks longer that the uncued half. The present experiment provides a direct test as to whether attentional cueing affects size processing in a visuomotor context.

## MO/14

MOVEMENT EXECUTION IN BLIND AND MENTALLY RETARDED. *D. Rapaic[1], J. Ivanus[2], G. Nedovic[1], V.Piscevic[2]. [1] Faculty og Defectology, University of Belgrade, Visokog Stevana 2. 11000 Belgrade, Yugoslavia. [2] University Center for Multidisciplinary Studies, Kneza Viseslava 1. 11000 Belgrade, Yugoslavia*

Objective of study: Voluntary "simple" and "complex" movements in blind and mentally retarded subjects of both sexes were studied in order to estimate background and capacity of the handicap. 15 were 16 years and another 15 were 19 years of age, without neurologic deficit. Method of used: They were tested by the Protocol for praxis estimation and scored as (a) correct, (b) movements with errors and (c) absence of movements.

Results: Matching results obtained gave nine categories – types of errors could be distinguished: executive, conceptual, body part as an object, uncritical to distance or object, omission of sequence, inversion of sequence, inadequvate gripping, sequence addition and substitution. Among them conceptual errors dominate up to 32% in mentally retarded, while in blind it rised up to 40,85%. Conclusions: These findings indicate the importance of cognitive psychology tools implementation into research of humane handicap sciences.

## MO/15

BIMANUAL FORCE COORDINATION: REDUCTION OF EEG NEGATIVITY IN PATIENTS WITH CEREBELLAR ATROPHY. *R. Verleger, E. Wascher, B. Wauschkuhn, P. Jaskowski, D. Kömpf & K. Wessel. Department of Neurology, Medical University of Lübeck, D 23538 Lübeck, Germany*

Lesions of the cerebellum cause disturbances of movement coordination. We developed an environment where this deficit could be measured and investigated in detail. Participants had to exert graded levels of force with one or two hands, and the time courses of force and of EEG potentials before and during movement were recorded. 12 patients with cerebellar atrophy and 10 healthy participants were studied. As could be expected, the patients had least success in the task where right and left hand had to exert different forces at the same time. This failure could be measured as the final deviation from the target force, but was also already seen at movement onset, by the larger difference of start time between both hands and by the increase of force which was approximately equal for both hands in the patients but proportional to the different target forces in the control group. A large, strongly Cz-focused negativity was seen in the control group, both during preparation for the movement and during movement. This negativity was larger in the more difficult tasks. In the patients, this negativity was smaller in all tasks, and the Cz focus was missing. This lack of negativity may reflect a deficit of activation of the motor areas. The undifferentiated reduction of negativity in all tasks is in contrast to the patients' specific failure in the most difficult task. This contrast suggests that their ever-present deficit of motor activation is still sufficient to perform in easier tasks but not in the task requiring difficult coordination. More general, the role of the cerebellum in movement coordination might be to selectively and differentially activate the motor cortices.

Supported by the Deutsche Forschungsgemeinschaft (Ve 110/5-1)

## MO/16

ANTICIPATORY SILENT PERIOD PRECEDING SELF-GENERATED OR -TRIGGERED RAPID TAPS – FEEDFORWARD CONTROL OF PROPRIOCEPTIVE FEEDBACK? *A. Struppler and P. Havel. Motor Control Research Unit, Klinikum r.d. Isar, Technical University Munich, Ismaninger StraÔe 22, 81675 Munich, Germany*

A brief hammer tap applied actively by a subject to the radial region of the opposite forearm is preceded by a dramatical reduction in the electro-myogram (EMG) of the muscles of the target arm. This anticipatory silent period (ASP) could be obtained over all muscles acting around the elbow joint, most apparent in the forearm flexor muscles. The phenomenon could also be observed in the lower extremity, e.g. the knee extensors and the planta flexors. Anticipatory EMG reduction occurs even when the active tapping movement is unexpectedly arrested before the hammer hits the target forearm. However if the impact is frequently arrested the ASP disappears. In our studies we apply the tap by a torque motor, which is triggerd by the subject. The ASP starts 40-100 ms prior to the impact, it is independent of the tap-force and lasts until the appearance of the short-latency reflex response.

Conditions for the occurence of ASP are the self-triggerd dynamic tapping in the intrapersonal space, the expected blow on the target forearm and a shorter delay than 20 ms between the expected impact and the active tapping. In this bimanuel action the command for the tapping movement and the command for the inhibition of tonic activity are precisely coordinated. It is likely that the preprogramming of the command for ASP is based on the experience of preceding events and the associated memory information received by processing of the sensory feedback from muscles, joints and skin.

In contrast to the wellknown anticipatory actions for stabilization of the limb the ASP does not counteract the perturbation. Increased viscoelastic absorbtion of the impact by decreased muscle stiffness seems also not to be the function because the main resistance to the tap is the forearm inertia. As proprioceptive afferents supporting tonic activity seem necessary for the occurence of the ASP, we investigated patients suffering from deficiencies in proprioceptive afferents on dorsal root level and following nerve repair. All patients had no T-reflex response and only a negligeable weakness in the elbow flexors. On the affected side, the ASP was considerably diminished.

We suppose that the ASP is involved in an feed-forward control of sensorimotor afferents on spinal or receptor level in order to reduce self-exaggerated afferent bursts caused by the self-generated tap already on mechano-receptor level.

## MO/17

IS THE ORGANIZATION OF GOAL-DIRECTED ACTION MODALITY-SPECIFIC? TEMPORAL ASPECTS. *P. Weiss[+], Y. Paulignan, M. Jeannerod, H.-J. Freund[+]. INSERM U94, Vision et Motricité, 16 av Doyen Lepine, 69500 Bron, France. [+] Department of Neurology, Heinrich-Heine-University, 40225 Düsseldorf, Germany*

Goal-directed action is triggered by sensory input from different modalities. While for simple movements the influence of different sensory input on movement organization is well described, comparable data about complex action is lacking. Since disturbances of complex action organization as in the apraxias are shown to be modality dependent, the effect of different input modalities on the temporal organization of the complex daily activity of 'Drinking from a bottle with a glass' was examined with the help of kinematic analysis.

For this purpose, this complex action was recorded in three-dimensional space with the help of an optoelectronic device in 12 normal subjects under four conditions: (1) action pantomime after verbal instruction, (2) action imitation after observation of the action performed by the experimenter without the objects, (3) action pantomime while seeing, but not touching the objects, and finally (4) action execution with objects.

In spite of high execution variability, the temporal structure of the complex action could be precisely described by the relative duration and peak velocity of action segments. In addition, the grip aperture-object size-correlation was stable across modalities. Interval and linear regression analysis between the onsets of functionally related action segments revealed functional bimanual coordination, which was achieved independent of modality. Therefore, the structure of normal complex action characterized by these kinematic parameters was not influenced by different instruction modalities. Only when the action was executed with the objects, did kinematic changes occur in the form of a reduced interval between the movement onsets of either hand and reduced peak velocity of the manipulative acts, while again no change was observed across the other three instruction modalities.

The fact that the temporal action structure, the grip aperture-object size-relationship, and the functional bimanual coordination were stable across instruction modalities suggests that a common (supramodal) action representation is accessed by different instruction routes.

# Neglect

## N/1
KINEMATICS OF REACHING AND GRASPING OBJECTS IN NEGLECT PATIENTS: *A. Farnè[1], A Roy[1], Y. Paulignan[1], G. Rode[2], D. Boisson[2], Y. Rossetti[1] and M. Jeannerod[1]: [1] Vision et Motricité, Inserm U94, Bron 69500, France; [2] Service de Rééducation Neurologique, Hopital Henry Gabrielle, Saint Genis Laval*

Neglect is generally defined as a deficit in which patients fail to orient, report and respond to information coming from the left, contralesional side of space (Heilman et al., 1993). Despite the fact that a great amount of literature devoted to this syndrome has revealed disturbances in the execution of exploratory movements, a kinematic analysis of reach to grasp movement in the peripersonal space is still vacant. In this study we present the first description of spatial and temporal characteristics of prehension movements performed by patients with unilateral neglect as assessed by classical clinical tests. Besides providing a basic knowledge of how neglect may influence the reach and the grasp components of object-oriented action, we address a second important issue concerning the effects of changing object position (Paulignan et al., 1991). This allows to clarify the residual ability of neglect patients to correct arm trajectory during movement in order to respond to a sudden change of object's visual location in space.

Five translucent, cylindrical objects (1.5 cm in diameter, 10 cm height) were presented to subjects. There were positioned on a table concentrically to the subject's body axis, from 20 degrees to the left to 20 degrees to the right. Under each object a red LED was embedded in the table. The objects were then illuminated, one at a time, by illuminating the corresponding LED. The subjects were instructed to reach and grasp the illuminated object. Two conditions were used, one where the subject had to grasp one of the objects randomly presented on each position (fixed condition). For the second condition, in 78% the central object was illuminated and the subjects grasped it. In the remaining 22%, perturbation of object location were applied at the movement onset. The amplitudes of the perturbations were of ten or twenty degrees to the right or to the left of the central target. The movements were recorded by mean of an optoelectronic motion analysis system OPTOTRAK 3020. The displacement of the thumb, the index and the wrist were sampled 200 times per second with a spatial precision of 0.1 mm.

Two main results were obtained. One concerns the neglect and the second one the lesion site. The patients were reaching and grasping normally fixed objects even in the left side of space. The objects were not neglected neither in a grasping nor in a verbal task. But in the perturbed condition, when the object location briskly changed, an asymmetry appeared on time landmarks. The second result shows different effects depending on the lesion site. When the frontal lobe was damaged deficits mainly appeared in the planning of action, measured by the reaction time. When the parietal lobe was affected, kinematics parameters were mainly disturbed.

Then to observe the classical asymmetry of a neglect patient behavior it is necessary to use a complex task such as a perturbed prehension movement. Conversely grasping a stationary object, taking into account all the object's properties, is not affected by a neglect syndrome if the subject is able to detect the target.

## N/2
DISPLACEMENT OF SUBJECTIVE BODY ORIENTATION IN NEGLECT PATIENTS WITH VS. WITHOUT HEMIANOPIA AND IN PATIENTS WITH HEMIANOPIA WITHOUT NEGLECT. *S. Ferber & H.-O. Karnath. Department of Neurology, University of Tübingen, Hoppe-Seyler-Str. 3, D-72076 Tübingen, Germany*

Recent studies argued for an altered representation of egocentric space in neglect patients with right parietal lesions which manifests itself in a deviation of the spatial reference frame toward the ipsilesional side. It is an ongoing discussion whether or not a rightward displacement of perceived straight ahead body orientation can be regarded as an inherent component of neglect. This debate is partly based on findings that reported the influence of other factors than pure neglect on the perception of 'straight ahead'. It has long been known that visual field defects like hemianopia, e.g., have an impact on the perception of body orientation. These findings were usually based on line bisection tasks which are a rather indirect and critical measure of the perceived orientation of the body in space. The aim of the present study was to disentangle the influence of primary visual field defects and the clinical manifestation of neglect on the perception of 'straight ahead'. We investigated neglect patients with and without hemianopia and compared them to patients with pure left hemianopia without neglect. In complete darkness, patients had to verbally direct a red LED – presented at eye level – to a position which they felt to lie exactly straight ahead of their bodies' orientation. We found a disparity of objective and subjective body orientation that was to the ipsilesional side in patients with pure neglect and to the contralesional side in patients with pure hemianopia. This disparity was less pronounced when both clinical symptoms were present, indicating that a combination of both effects might neutralize the misperception of body orientation caused by each disorder individually.

## N/3
DIFFERENTIAL RECOVERY PATTERNS IN TWO PATIENTS WITH HEMISPATIAL NEGLECT. *Harvey, A. Rowan, H. Miller. Department of Psychology, University of Bristol, Bristol, BS8 1TN, England, U.K.*

A number of recent studies have shown that there are neglect patients with a broadly-defined perceptual-type neglect who demonstrate an attentional/representational failure resulting in a misperception of space, whereas other patients suffer from neglect of a mainly premotor nature thereby failing to direct their actions properly (Bisiach et al., 1990, Tegner and Levander, 1991).

The easiest way to distinguish between these two broad alternatives is the Landmark task, in which the patient is asked to indicate which end of the line a centrally bisected mark is closer to. Milner and Harvey (1995) have shown that perceptual-type neglect patients experience a distortion of their subjective space so that the central bisection mark appears to be offset to the left. This subjective distortion proves more marked for lines presented in left than central and right hemispace and shows that any potential premotor biases are outweighed by the subjectively experienced distortion.

The current experiment addresses the question of a potential recovery from these perceptual biases. Two patients with hemispatial neglect were tested on two occasions (3 and 21 months post stroke) with both line bisection and landmark tasks. During the first testing session both patients showed strong TperceptualU neglect with large rightward errors in the standard bisection task (the patient is asked to indicate the centre of the line) and predominantly leftward pointing in the landmark test. On the second occasion, however, both patients showed a marked recovery when tested with the line bisection task in that they displayed only small right or leftward errors. In contrast, their landmark performance was still markedly affected with almost 100% leftward pointing, although this tendency proved slightly less marked in right hemispace.

These data suggest that both patients still experienced a subjective distortion of their visual space, whereas any putative response biases that may have originally contributed to their impaired bisection performance had clearly recovered. This suggests rather differential recovery patterns for the two neglect types, with the perceptual bias as the more resistant to recovery. It also shows that the landmark test is a sensitive means to identify these remaining abnormalities.

## References
Bisiach, E., Geminiani, G., Berti, A., and Rusconi, M.L. (1990). Perceptual and premotor factors in unilateral neglect. Neurology ,40, 1278–1281.

Harvey, M., Milner, A.D., and Roberts, R.C. (1995)). An investigation of hemispatial neglect using the landmark task. Brain and Cognition, 27, 59–78.

Tegner, R., Levander, M. (1991). Through a looking glass. A new technique to demonstrate directional hypokinesia in unilateral neglect. Brain, 114, 1943–1961.

## N/4

DISENTANGELING GRAVITATIONAL AND EGOCENTRIC COORDINATES IN SPATIAL NEGLECT: *H.-O. Karnath, M. Fetter & M. Niemeier: Department of Neurology, University of Tübingen, Hoppe-Seyler-Str. 3, D-72076 Tübingen, Germany*

Rotation of the head and/or body to the left and to the right around the roll-axis revealed that visual stimuli were neglected on the egocentric, body-centered left but also on the gravitational left of neglect patients. By using visual stimuli, these studies had to confound the gravitational left with the left side of a visual display. However, visual stimuli are known to intrinsically influence the patients' perception and behaviour. Therefore, in the present experiment we evaluated the influence of gravity on contralateral neglect when no visual stimuli were presented. A technique which serves for this purpose is the observation of exploratory eye movements while searching for a non-existent target in complete darkness. Neglect patients show a bias of exploratory eye movements toward the ipsilesional side under this condition. The influence of gravity on the patients' exploratory eye movements was investigated in five experimental conditions: (I) body in normal upright position, (II) body tilted 30° to the left and (III) 30° to the right, (IV) body pitched 30° backward and (V) 30° forward. A covariance ellipse was computed on the eye movements. The orientation of its axes was taken as a measure for the orientation of the visual search field. In body coordinates, we found no significant differences between the orientation of the visual search fields in the different conditions suggesting that the search field shifted with the orientation of the body. A slight but statistically not significant tendency to diminish neglect (here by an extension of visual search toward the contralesional left) was observed when the body was pitched backward. In conclusion, the present results revealed that the patients' failure to explore the contralesional part of space is related to an egocentric, body-centered reference frame and that gravity has no specific influence on the exploratory bias of these patients.

## N/5

DISORDERS OF VISUOSPATIAL PERCEPTION IN THE ROLL PLANE IN PATIENTS WITH NEGLECT: *G. Kerkhoff, C. Zoelch: EKN- Clinical Neuropsychology Research Group, Krankenhaus Bogenhausen, Dachauerstr. 164, D-80992 München, Germany*

Patients with hemispatial neglect fail to detect or respond to visual, acoustic or tactile stimuli in their contralesional hemispace. Recent models of spatial neglect have interpreted these disturbances as evidence for a systematic shift of an egocentric coordinate system towards the ipsilesional hemisphere. However, neglect patients often show visuospatial deficits, i.e. disturbed perception of orientation discrimination and impaired judgment of the Subjective Visual Vertical (SVV) and Horizontal (SVH) which can not be explained within such a model. Other evidence suggests that patients with parieto-insular vestibular cortex lesions show disturbed perception of the (SVV) although the relationship to visual neglect was not analyzed. We therefore investigated if patients with visual neglect show visuospatial deficits in the roll plane, by measuring their judgment of the SVV, SVH and visual discrimination of oblique orientations. Thirteen patients with right hemispheric vascular lesions and *left spatial neglect* documented by clinical tests (number cancellation, object search, line bisection, drawing), 14 patients with right hemispheric vascular lesions *without left spatial neglect* and 12 *normal control* subjects were tested in a PC-based measurement of the SVV, SVH and a 45 oblique orientation discrimination task. All subjects performed 10 trials in each of the three tasks, five with a clockwise and five with a counterclockwise rotation towards the gravitational vertical (step-width 0.5, size of the bar: 16 cm × 1.4 cm). The head and body were oriented earth-vertical within an experimental chair and a head- and chinrest.

Normal subjects showed a nearly perfect judgment of the SVV (defined as 90), SVH (defined as 180) and the 45 orientation discrimination task (SVV: M 89.6, sd 0.8; SVH: M 179.6, sd 0.4: 45: M 44.9, sd 1.4). Neglect patients showed significant deficits in all three tasks: SVV: M 94.9, sd 3.8; SVH: M 183.7, sd 2.8; 45 M 49.7, sd 4.5). Patients without neglect did not show deficits in any of these three tasks (SVV: M 89.8, sd 0.5; SVH: M 179.8, sd 0.5; 45 M 45.3, sd 2.3). Analysis of variance revealed that neglect patients were significantly impaired in all three spatial orientation task as compared to control subjects and patients without neglect whereas the control patients performed like normal controls.

These data show that neglect patients show visuospatial deficits that cannot be explained by a systematic, ipsilesional deviation of an egocentric coordinate frame. Rather, they indicate deficits in spatial orientation and a contraversive shift of visual space in the sagittal plane. These deficits are evidence in favour of a disturbed egocentric coordinate system not only in the horizontal but also in the roll plane in patients with visual neglect.

## N/6

THE INFLUENCE OF VISUAL PERCEPTUAL BIAS ON POINTING RESPONSES TO AUDITORY STIMULI. *E. Làdavas[1], F. Pavani[2]. [1] Department of Psychology, University of Bologna, viale Berti Pichat 5, 40127, Bologna, Italy. [2] Hospital "I Fraticini" IN-RCA, via dei Massoni 22, 50100, Firenze, Italy*

A previous study has shown that a right biased visual representation in neglect patients may influence the ability to judge the subjective midline (Farné et al. 1997).The aim of the present study is twofold. The first one is to verify whether this impaired visual representation can be modulated by an intact proprioceptive representation related to the position of response effectors. The second one is to evaluate whether the impaired visual representation influences a localization performance in the auditory modality. To these aims, a patient (G.A.), with visual neglect and hemianopia, was tested in partial vision (Exp. 1) or blindfolded condition (Exp. 2). In the first experiment, only the vision of the hand was prevented by the use of a cardboard placed on the top of the hands. One single auditory stimulus was presented on the left, centre or right space. The patient was required to point the location of the sound with the left or right hand, each located on the left, centre or right space with respect to the body midline.The accuracy of response was measured on a semicircular scale. The performance of the patient was compared with that of control subjects matched for age and years of schooling.

The results of Exp. 1 showed a rightwards bias in the auditory localization task, since all the manual responses were confined to the right space. In addition, this bias was modulated by the location of the response effectors in the space. It was found a decrement of performance when both effectors were located on the right side, whereas an improvement was found when they were located on the left side of the body midline. However, the pathological bias found in Exp. 1 was reduced in Exp. 2, in which the patient was blindfolded, and therefore no visual information was available.

In conclusion, the results of the present study showed a modulation of visual neglect determined by the proprioceptive information about the location of the responding effectors, and an influence of the impaired visual representation on the auditory localization task.

### References

Farné, A., Ponti, F. & Làdavas, E. (1997) In search for biased egocentric reference frames in neglect. Submitted

**N/7**

VISUAL SEARCH IN THE ENVIRONMENT BY PATIENTS WITH NEGLECT: *M. Niemeier & H.-O. Karnath: Department of Neurology, University of Tübingen, Hoppe-Seyler-Str. 3, D-72076 Tübingen, Germany*

Clinical tests for neglect, like the letter cancellation task, frequently show that neglect patients not only ignore the target stimuli on the contralesional side but also tend to repeatedly cancel the same target stimuli on the ipsilesional side. This observation matches with Kinsbourne's attentional model of neglect which postulates a bias of attentional orienting to the ipsilesional side leading to a gradient along the horizontal axis with a hypoattented contralateral side and a hyperattented ipsilesional side. However, clinical tests (as well as the display of a PC monitor) usually examine visual search only in a small area of approximately +/− 20° left and right of sagittal midplane. The aim of the present study was to study visual search of neglect patients in a more natural situation. A random configuration of letters was presented on the inner surface of a sphere that surrounded the subject. The subjects were sitting in the center of the sphere and were requested to search for a single target letter 'A' that was stated to be presented 'somewhere in the sphere'. In fact, the target was not presented while exploratory eye movements were recorded. As in controls, the distribution of exploration along the horizontal axis was roughly bell-shaped. In contrast to controls, the whole search field was deviated to the ipsilesional side. In conclusion, neglect patients demonstrated no hyperattention of the extreme ipsilesional side by orienting gaze in that direction and spending most of the time searching there. The results rather favour the assumption of a deviated egocentric reference system in patients with neglect.

**N/8**

Contribution to the 29th Annual General Meeting of the European Brain and Behaviour Society (EBBS), September 15–18, 1997 in Tutzing/Bavaria/Germany

MOTOR NEGLECT OR SIMPLY SHIFT OF HAND PREFERENCE? NEUROPSYCHOLOGICAL DIFFERENTIAL DIAGNOSIS AIDED BY TRANSCRANIAL MAGNETIC STIMULATION (TMS) AND KINEMATOGRAPHY IN A CASE OF GERSTMANŃS SYNDROME. *F. Uhl[1,2], J. Hermsdörfer[1], G. Kerkhoff[1], G. Goldenberg[1]. [1] Entwicklungsgruppe Klinische Neuropsychologie, Krankenhaus München-Bogenhausen, Dachauerstr. 164, D-80992 München and [2] Neurologische Univ.-klinik, A-1090 Wien (Director: Prof. L. Deecke)*

A 65 year old accountant with left parietal stroke demonstrated not only Gerstmanńs syndrome and ideomotor apraxia, but also underutilization of the right hand:
E.g. when reaching for an object in right hemispace, he spontaneously used his left hand, instead of the adjacent right hand. After having completed grasping, he became aware of the peculiarity of his action, blaming himself for having "again neglected the right arm". Also, underutilization of the right arm was obvious (ii) in spontaneous gesticulation during conversation, (iii) during gait, (iv) in using only the left hand during HAWIE-mosaic-test, with the right hand resting passively on the table.
However, for writing and drawing, he still used the right hand. There was no functionally-relevant paresis nor motor extinction or motor impersistence. Rapid alternating movements, such as index finger tapping were slower at the affected right hand.
*Neuropsychological differential diagnosis:* Such underutilization of the right arm could be interpreted as motor neglect or as (involuntary) shift of hand preference towards the left hand. The latter would result from the combination of three factors: the loss of dexterity at the affected right hand, together with the apraxia and the severe right-left body scheme disorientation.
To answer this question, two methods were applied.
I) In *Transcranial magnetic stimulation* (TMS), a prolonged silent period (SP) followed the motor evoked potential (MEP). SP was twice as long for the affected left hemisphere (400 msec) than for the right hemisphere (200 ms). This points to exaggereated intracortical inhibitory processes (according to R. Benecke's group: Classen et al., Brain, in press). They in turn favour the diagnosis of motor neglect.
II) In *Kinematography* of finger tapping, the reduced frequency at the affected right hand turned out to result from covert pauses following completion of an individual movement cycle. Such pauses correspond to a disturbance of movement (re-)initiation. Again, this is in favour of the differential diagnosis of motor neglect.

**N/9**

DISSOCIATION BETWEEN FAR AND NEAR VISUAL SPACE IN NEGLECT. *P. Vuilleumier[1], N. Valenza[1], E. Mayer[1], A. Reverdin[2], T. Landis[1]. Department of Neurology and Neuropsychology[1]; and Department of Neurosurgery[2]; University Cantonal Hospital of Geneva, 24 rue Micheli-du-Crest, 1211 Geneva, Switzerland*

Background: The recent report [Halligan and Marshall, Nature 1991; 350: 498–500] of a patient with left visual neglect for near but not far space after a right parietal stroke suggested that the human brain might have separate attentional mechanisms for far (extrapersonal) and near (peripersonal) space. Studies in monkeys have provided substantial evidence for such a modular organization. However, other studies of neglect patients failed to find a comparable dissociation and the inverse dissociation, a selective inability to attend to far but not near space due to a differently placed lesion, is needed to support this claim. Methods: We now describe a patient with severe unilateral left visual neglect in far-distant space but not within near reaching space following a right temporo-occipital heamatoma. Her performance in a series of neglect tasks was systematically compared in near and far distance conditions, with the visual angle of stimuli being kept constant across conditions. In tasks that required manual responses, the patient used a pencil in near space and a laser pointer in far space. Results: Left visual neglect was consistently observed in far but not near space for all perceptual (e.g. reading, dots counting, square completion) and perceptual-motor tasks (e.g. letter and star cancellation, line bisection) and was most important in the latter. Furthermore, though she made rightward errors when indicating the midpoint of far-distant lines, our patient made smaller opposite leftward errors for near-distant lines. Conclusion: This case provides a definitive support for a modular representation of far and near visual space in humans. It also supports a predominant involvement of inferior temporo-occipital cortical areas, hiterto mainly associated with form and colour processing, in the control of attention towards extrapersonal visual space.

**N/10**

ILLUSORY CONTOURS AND NEGLECT. *P. Vuilleumier & T. Landis. Department of Neurology, University Hospital of Geneva, 24 rue Micheli-du-Crest, 1211 Geneva, Switzerland*

Objective: Perception of illusory contours was investigated in neglect patients with the hypothesis that visual grouping processes operate on a preattentive basis and that the detection of illusory figures might be dissociated from the overt detection, or awareness, of their lateral edges. Methods: The same illusory lines and rectangles were presented in two conditions (1) an implicit detection task where patients had to mark the midpoint of given figures; (2) an explicit detection task where patients made same-different judgments for pairs of figures that could be similar or differ either by their right or left edge. In two further control tasks, patients were asked (1) to bisect full lines and full squares of same size as the previous lines and squares; (2) to mark the midpoint between two vertical lines with comparable extent and interval as the previous lines and squares. Results and conclusion: Preliminary results in four patients with left visuospatial neglect indicate that (1) implicit and explicit perception of illusory contours can be dissociated in neglect, consistently with our hypothesis; (2) bisection of illusory figures can show a rightward bias similar to that of complete figures; (3) in some patients, implicit detection can lead to an explicit judgment of difference for pairs of dissimilar figures, with the left-sided differences being however neglected while irrelevant right-sided differences being reported.

# Psychopharmacology

## PP/1

ACAMPROSATE REDUCES OPERANT RESPONDING FOR ETHANOL MORE EFFECTIVELY DURING REINSTATEMENT THAN DURING BASAL DRINKING: *S.M. Hölter, W. Zieglgänsberger, R. Spanagel. Max-Planck Institute of Psychiatry, Kraepelinstr. 2, D-80804 München*

We studied the effects of the putative alcohol anti-craving compound acamprosate on operant responding for ethanol during basal drinking and after an ethanol deprivation episode in long-term ethanol experienced rats. All animals had the free choice of water, 5%, 10% and 20% ethanol solutions for five months before the start of operant sessions. Then the animals were divided into two groups: the group "basal drinking" continued to have access to all ethanol solutions until operant testing, whereas the group "reinstatement drinking" was tested in the operant chambers only after one week of ethanol deprivation. The operant test was a two-lever free-choice paradigm with 20% ethanol and concurrent water. The reinstatement drinking group exhibited a strong alcohol deprivation effect with immediate high ethanol consumption (approx. 2 g/kg during the first hour) and ethanol preference (72%). Acamprosate (100, 200 and 400 mg/kg) dose-dependently reduced ethanol consumption and preference in the reinstatement drinking group during the first hour of testing. However, in the basal drinking group only the two higher doses of acamprosate were effective.

We conclude that (a) acamprosate interferes with the reinforcing properties of ethanol and (b) its increased efficacy during reinstatement drinking indicates an anti-craving effect.

## PP/2

SPATIAL LEARNING DEFICITS IN MICE WITH A TARGETED GLUCO-CORTICOID RECEPTOR GENE DISRUPTION. *M.S. Oitzl, O.C. Meyer[3], T.J. Cole[2,3], W. Schmid[2], G. Schütz[2], M. Joels[1] and E.R. de Kloet. Division of Medical Pharmacology, Leiden/Amsterdam Center for Drug Research, Leiden University, P.O.Box 9503, 2300 RA Leiden, and[1] Department of Experimental Zoology, University of Amsterdam, The Netherlands.[2] Division of Molecular Biology of the Cell I, German Cancer Research Center, Heidelberg, Germany.[3] current address: OCM – Department of Physiology, University of California, San Francisco CA94143-0444, USA; TJC – Baker Medical Research Institute, Prahran, Victoria, Australia.}*

Previous studies in rats using the Morris water maze suggested that the processing of spatial information is modulated by corticosteroid hormones through mineralo- and glucocorticoid receptors (MRs and GRs) in the hippocampus. MRs appeared involved in the modulation of explorative behavior, while additional activation of GRs facilitated the storage of information.

In the present study we have examined spatial learning and memory of mice homozygous and heterozygous for a targeted disruption of the GR gene in the water maze. Compared to wild type controls, homozygous and heterozygous mice were impaired in the processing of spatial but not visual information. Homozygous mutants performed variably during training, without specific platform directed search strategies. The spatial learning disability was partly compensated by increased motor activity. The deficits are indicative for a dysfunction of GRs as well as MRs. In contrast to heterozygous and wild type animals, homozygous mice displayed also an increased locomotor activity in the open field. Heterozygous mice performed comparable to wild type mice with respect to the latencies to find the platform. However, their strategy was more >comparable to the homozygous mice. GR-related long-term memory was found to be impaired. These mutants displayed an increased behavioral reactivity in the open field, which points to a more prominent MR-mediated function. In line with this, in situ hybridization studies performed in the same group of animals showed that MR mRNA expression in CA2 and CA3 hippocampal areas of heterozygous mutants is increased compared to the homozygous mice. In the hippocampus of homozygous mutants GR mRNA was undetectable,

whereas the heterozygous animals showed a specific reduction of GR mRNA in the CA1 area, the functionality of which is strongly correlated to the performance of spatial tasks in rats and man. Basal plasma levels of corticosterone were high in homozygous and intermediate in heterozygous when compared to wild type mice.

At present, the GR-knockout mice represent the only available animal model in which the effects mediated by MR can be studied in the complete absence of GR. The findings indicate that (i) the GR is of critical importance for the control of spatial behavioral functions, and (ii) MR-mediated effects on this behavior require interaction with functional GRs. This allowed us to propose that for these hippocampal-related MR functions in behavior interaction with GR is required.

Supported by the Dutch Organization for Scientific Research NWO 554-545 and EC Biotechnology Programme PL960179.

## PP/3

EFFECTS OF ENVIRONMENTAL CONDITIONING ON THE EXPRESSION OF MORPHINE-INDUCED MESOLIMBIC SENSITIZATION. *Inge Sillaber, Philipp Kämpf, Rainer Landgraf and Rainer Spanagel. Max Planck Institute of Psychiatry, Clinical Institut, Department of Neuroendocrinology, Kraepelinstr. 2, 80804 Munich, Germany*

Classically conditioned factors play an important role in addictive processes and drug relapse.

In the present study we examined the influence of an environmental stimulus upon sensitized mesolimbic dopamine (DA) release induced by repeated intermittent injections of morphine. Rats were pretreated for 7 days with morphine (10 mg/kg, s.c.) or saline; injections were made either in the test environment (paired) or in the homecage (unpaired). Three days after the pretreatment procedure the animals were challenged with a 10 mg/kg morphine dose. Using the in vivo microdialysis technique, we found that the extracellular dopamine concentration in the nucleus accumbens was significantly enhanced when drug pretreatment was paired with the external stimulus, in contrast to the unpaired condition. As behavioral sensitization has also been shown to become influenced by external cues, these results underline the important role of conditioning in the process of sensitization and its mediation via the mesolimbic DAergic system.

## PP/4

AGE – DEPENDENT BEHAVIOR PATTERNS OBSERVED AFTER PHENCYCLIDINE ADMINISTRATION. *R. Veskov, Z. Ostojic, M. Car. G. Konjevic, S. Ruzdijic, Lj. Rakic. Department of Neurobiology. Institute for biological research, 11060 Belgrade, Yugoslavia*

Phencyclidine (PCP) is non-competitive NMDA antagonist, a dopamine and norepinephrine uptake inhibitor, and also has high affinity for the sigma binding site. Having in mind that aging produces a general decline in the efficacy of synaptic transmission in central nervous system, we presume that characteristic behavioral sets induced by PCP in animals are quite different and age dependent. This hypothesis was tested on 3-, 12- and 24-month – old male Mill Hill hooded rats, assigned randomly as control and drug treated groups of 8 animals. They were injected with single dose of 5, 10 or 50 mg/kg of PCP or physiological solution. Each animal was tested only once in computerized infrared beam system with 16 IR beams in each axis. The gross behavior including locomotor, sterotypic, ataxic and seizure activity were recorded 120 min continuosly. Ambulatory activity were analyzed by Student's test and two way ANOVA. Stereotypy and ataxia were measured using the scale described by Sturgeon, R.D. et al. 1979. There was a significant differences in behavioral effects after PCP injection between young and aged rats. Our finding suggest that PCP in young rats caused hyperlocomotion (3 month old), ataxia and stereotyped behaviors (sniffing, head-weaving, backpedalling and turning) at doses 5 or 10 mg/kg. In the same time ambulatory movements of old animals (12 month, the same results were observed in 24 month old rats) were less increased, while their stereotypic movements and ataxia were significantly increased in comparison with younger rats. Behavioral seizures consisting of single myoclonic jerks were more prominent in young rats. Application of higher doses (50 mg/kg) of PCP decreased ambulatory activity and animals fall on the floor of testing chamber in characteristic position. Frequent myoclonic jerks were seen in young rats and lasted more than 90 min. Convulsive activity in aged rats was observed very rarely.

Our results indicate that manifestation of characteristic components of PCP behavior pattern of the rat is dose and age dependent. These defferences result from specific changes of neurotransmitter systems in the brain during development and aging.

# Time

## T/1

TIMING THE ONSET OF MOVING AND STATIONARY STI-
MULI: *G. Aschersleben, J. Müsseler, L. Knuf & W. Prinz. Max
Planck Institute for Psychological Research, Leopoldstraße 24; D-
80802 München; Federal Republic of Germany*

When participants are asked to localize the first position of a moving
stimulus they typically mislocalize it in the direction of the movement (Fröhlich Effect; Fröhlich, 1923). As possible mechanisms
causing the Fröhlich Effect a low-level motion-deblurring mechanism and a high-level attentional account are discussed (Aschersleben & Müsseler, 1997; Müsseler & Aschersleben, 1996). In anyway,
the mislocalization points to a temporal error indicating a delay in
the subjective timing of a moving stimulus. However, this delay is
in contrast to other findings according to which moving stimuli
are processed faster than stationary stimuli. We explored this dissociation in four experiments.
In Experiment 1 the effect was established spatially. Asked to judge
the starting position of a fast moving stimulus, participants tended to
make localization errors in the direction of the movement as compared to the judging of a stationary stimulus flash. In the subsequent
three experiments three different temporal tasks were examined under otherwise identical conditions: simple reaction times, temporal
order judgments, and synchronization performances. Reaction times
were shorter to the onset of a movement as compared to the onset of
a flashed stimulus (Experiment 2). Conversely, the other experiments revealed a delay in the perceived timing. The moving stimulus had to be presented earlier than the flashed stimulus to be judged
as being simultaneous with a comparison click, that is, the perceived
onset of the moving stimulus was delayed as compared to the perceived onset of the flashed stimulus (Experiment 3). Finally, in a
synchronization task the size of the asynchrony between the pacing
signal and the motor response was increased with the flashed stimulus as compared to the moving stimulus indicating that the timing of
the motor response was delayed when the moving stimulus served as
the pacing signal (Experiment 4).
In conclusion, the reported experiments give evidence for the idea
that the mislocalization at the beginning of a movement is not pri-
marily a spatial effect but rather a temporal one. Moreover, this finding indicates that a stimulus can affect motor responses and perceptual judgments in a different manner. The output of early stimulus
processing feeds directly into the motor system whereas the representation that is used for localization judgment, temporal order judgment, and synchronization performance is based on later integrative
processes. This interpretation is in line with the distinction recently
put forward by Goodale and Milner (1992). On the basis of neuropsychological studies they argued that the neural substrates of visual
perception are quite distinct from those underlying the visual control
of action.

Aschersleben, G. & Müsseler, J. (1997). Dissociations in the timing
of stationary and moving stimuli. Ms submitted for publication
Fröhlich, F. W. (1923). Über die Messung der Empfindungszeit.
Zeitschrift für Sinnesphysiologie, 54, 58–78
Goodale, M. A., & Milner, A. D. (1992). Separate visual pathways
for perception and action. Trends in Neuroscience, 15, 20–25
Müsseler, J. & Aschersleben, G. (1996). Zur Rolle visueller Auf-
merksamkeitsverlagerungen bei der Etablierung einer (subjektiv be-
richtbaren) Raumrepräsentation. In B. Mertsching (Ed.), Aktives Se-
hen in technischen und biologischen Systemen (pp. 83–92). Sankt
Augustin (FRG): Infix

## T/2

AGRAMMATISM FOLLOWING A CEREBELLAR LESION:
THE PROBLEM OF TEMPORAL INTEGRATION AND SYN-
CHRONIZATION IN SPONTANEOUS LANGUAGE PRODUC-
TION. *E.G. de Langen. Klinikum Passauer Wolf, Department of
Neurology, Postfach 1263, D-94083 Bad Griesbach, Germany*

A case of a 74-year old right handed patient, MB, with a right cerebellar hemorrhage is reported. She developed a Broca's aphasia with
agrammatism. Cognitive deficits following cerebellar lesions are not
unusual and aphasic disturbances are described in some cases, but
agrammatism is very rare and only one similar case is reported in
literature (Silveri et al, 1994). This case leads to the question, if
the functional interrelation between supratentorial structures and
the cerebellum causes agrammatism as a consequence of diaschisis
to higher cerebral functions or, if the cerebellum itself contributes as
a functional part in the language production system. In aphasiology
exists a chronogenetically tradition, which offers, in combination
with psycholinguistic and neurophysiological theories of language
production, some hypotheses about the relationship between higher
cortical functions and the cerebellar system.
The patient was investigated 5 weeks p.o. A mild Broca's aphasia
with a characteristic agrammatism was diagnosed. Neuropsychological assessment revealed an impairment of verbal short-term and
long-term memory and attention. Examination of motor performance showed a mild hemiataxia of gait and stance. MRI 9 weeks
p.o. showed a lesion in the right cerebellar hemisphere including
the dentate nucleus.
Modern theories of language production consider sentence production as an incremental procedure, processed serially within segments,
but in a parallel fashion between segments. This parallel processing
allows for the smooth flow of language in discourse. Kolk (1995)
states that for a correct production of a sentence, synchrony between
the various parts of the syntactic tree is necessary. In this theory of a
time-based approach to agrammatic production, the basic difficulty
that results from a timing deficit is that a particular representational
element decays before other elements, with which it has to be in synchrony, are activated. However, it is still unclear which mechanism
is responsible for the syntactic computation and the syntacto-lexical
integration.
Cerebellar structures are known to play an important role in the precise temporal interplay between different sets of muscles through the
cerebro-cerebellar loop. Pathology also demonstrates a cerebellar
participation in many human cognitive functions, especially in systems with temporally predictive computations. So, either the cerebellar function, strictly that of the dentate nucleus with well-known
projections to the thalamus and Broca's area, plays a specific role in
language production, or, there is a rather unspecific but crucial function of the cerebellum in oscillatory brain activity which coexists
with cognitive temporal binding in the thalamo-cortical system. A
cerebro-cerebellar loop is proposed to be responsible for a normal
function of the cortico-thalamo-cortical system in the production
of propositional spontaneous language.

## T/3

THE COGNITIVE REPRESENTATION OF TIME INVESTIGA-
TED WITH SEQUENTIAL MOVEMENTS. *M.H. Fischer. Depart-
ment of Psychology, University of Munich, Leopoldstr. 13, D-80802
Munich, Germany*

OBJECTIVE: Our cognitive representation of time is not well understood: Models of on-line movement control assume perfect knowledge of elapsed time (e.g., Meyer et al., 1988). Retrospective judgments of time intervals, however, reveal cognitive biases (e.g.,
Fraisse, 1984). Three experiments demonstrate that motor timing
is also prone to cognitive biases, and that time estimation is in turn
affected by motor demands.
METHOD: Participants alternated with their right hand between two
target locations at a prescribed rate. Movements were paced with a
metronome that was turned off prior to data collection. Targets were
$15 \times 15$ mm buttons placed 25 cm apart. Two time judgments were

obtained: (1) on-line judgments from movement times between targets, and (2) reproductions of estimated sequence completion times. In Experiment 1, participants completed movement sequences with either fast or slow speed. Targets were embedded in easy or difficult sequences. Participants also performed static tapping tasks and retrospectively reproduced their required sequence completion times in all tasks. In Experiment 2, short (3 vs. 6 second) sequences were used, and sequence completion times were estimated prospectively or retrospectively. Experiment 3 investigated how a physical load (453 g wrist weight) and a cognitive load (120 vs. 27 mm contact area) affected timing and time estimation.

RESULTS: In Experiment 1, movement times between identical targets were faster within more demanding sequences. Static tapping was equally accurate at all locations, indicating that the timing bias was not related to postural discomfort. Retrospective sequence duration estimates did not discriminate between conditions.

Experiment 2 replicated the anticipation of sequence difficulty and found overestimation of short and underestimation of long sequence completion times in all conditions. In Experiment 3, physical loading led to faster movements and raised the estimates of total sequence durations. Cognitive loading induced underestimation of sequence durations. Retrospective sequence duration estimates were shorter than prospective estimates for both short and long sequences, indicating that the motor activity itself, rather than memory decay, affected the represented sequence duration.

CONCLUSIONS: The present results suggest that moving and timing are primary vs. secondary tasks competing for cognitive resources (e.g., Thomas & Weaver, 1975). They also show that the widely held view in motor control research of an accurate time-keeper is mistaken: Cognitive biases similar to those in time estimation occur in the on-line timing of behavior, thus enabling a movement-based approach to understanding the cognitive representation of time.

REFERENCES

Fraisse, P. (1984). Perception and estimation of time. Annual Review of Psychology, 35, 1–36

Meyer, D. E., Abrams, R. A., Kornblum, S., Wright, C. E., & Smith, J. E. K. (1988). Optimality in human motor performance: Ideal control of rapid aimed movements. Psychological Review, 95, 340–370

Thomas, E. A., & Weaver, W. B. (1975). Cognitive processing and time perception. Perception & Psychophysics, 17, 363–367

## T/4

TEMPORAL DEGREES OF FREEDOM IN SENSORY-MOTOR COORDINATION. *R. E. Ganz[1] & W. H. Ehrenstein[2]. [1] Institut für Medizinische Psychologie, Universitätsklinikum Essen, Hufelandstr. 55, D-45122 Essen, Germany. [2] Institut für Arbeitsphysiologie, Ardeystr. 67, D-44139 Dortmund, Germany*

The complexity of timing in visuo-motor coordination was addressed by a method derived from the physics of non-linear dynamic systems. From Bernstein's conception of spatial degrees of freedom in motor coordination, we derive the correlation dimension, D, as a measure of the effective number of the temporal degrees of freedom that underly performance in a given sensory-motor task. The validity of the proposed estimator is demonstrated in two experiments on visuo-motor compatibility. Subjects moved a stylus on a linear rail to track a visually displayed target that also moved in a straight line. The target was presented with sinusoidal acceleration (frequency: 0.1 Hz; amplitude: 16 deg). Either the stylus track (Exp. I) or the trajectory of the target (Exp. II) was rotated in order to vary the degree of visuo-motor compatibility (in steps of 45 deg) between 'highly compatible' (angle between target and tracking directions: 0 deg) and 'highly incompatible' (180 deg). Time series measurements (2514 data points in intervals of 15 ms) of the spatial deviations of the tracking movements from the target trace ("error data") were made. For the computations of $\underline{D}$, a phase portrait of vectors $\vec{X}_{(t_i)}$ of the error data $x_{(t_i)}$ was reconstructed in an m-dimensional phase space such that $\vec{X}_{(t_i)} = [x_{(t_i)}, x_{(t_i+\tau)}, x_{(t_i+2\tau)}, \ldots x_{(t_i+(m-1)\tau)}]$, where $t_i$ denotes the discrete time with the index i running from 1 to the number

of data points, and $\tau$ is a preset time lag. The vectors $\vec{X}_{(t_i)}$ can be treated as though they represent the (multi-dimensional) temporal states of the visuo-motor system (Takens' theorem) and, thus, comprise what we call visuo-motor coordination. All pairs of vectors $\vec{X}_{(t_i)}, \vec{X}_{(t_j)}$ ($\tau = 0.1s$; m = 15; temporal distance of vectors $\vartheta > 1.5s$) within the Euclidian distance, $\varepsilon$, were summed according to the Grassberger-Procaccia algorithm to yield a correlation integral, $C(\varepsilon)$, which scales as $\varepsilon^D$. Thus, $\underline{D}$ was calculated as the slope of the regression line $\ln C(\varepsilon)$ against $\ln \varepsilon$ (scaling region: $-5 \leq \ln C(\varepsilon) \leq -4$). Visuo-motor coordination had a low-dimensional temporal structure with $\underline{D}$-values in the order of 6, which was consistent with deterministic chaos rather than with pure stochastic noise. $\underline{D}$ was closely related to the degree of visuo-motor compatibility (r = 0.95, p < 0.001 in Exp. I; r = 0.76, p < 0.05 in Exp. II) and was positively correlated with tracking performance (r = 0.58, p < 0.001 in Exp. I; r = 0.43, p < 0.001 in Exp. II): The smoother the tracking, the more temporal degrees of freedom were available or could be released for motor control. However, for short periods of training, tracking performance increased (p < 0.001 in Exp. I; p < 0.05 in Exp. II) whereas $\underline{D}$ did not (p > 0.50 in Exp. I and II). Further, the subjects who repeated the tracking task (compared to those without prior experience with the task) showed better performance (p < 0.05), but no significant difference in their respective values of $\underline{D}$ (p = .18). We conclude that the introduced measure of the number of the temporal degrees of freedom in visuo-motor coordination, $\underline{D}$, is sensitive to the dynamic complexity of motor control; more specifically, its reciprocal, $D^{-1}$, may reflect the solely task-related structure and costs of the underlying translatory mechanisms.

## T/5

ARE THE NEURAL NETWORKS INVOLVED IN TEMPORAL ESTIMATION DIFFERENT DEPENDING ON THE PROCEDURE USED? A CONTRIBUTION FROM PET STUDIES. *Lejeune H.[1], Maquet P.[2], Bonnet M.[3], Casini L.[3], Ferrara A.[1], Macar F.[3], Pouthas V.[4], Vidal F.[3]. [1] PTP Unit and [2] CRC, Liège, Belgium; [3] CNRS Marseille and [4] CNRS Paris, France*

### Introduction

We previously showed that the estimation of very short durations, between 0.5 and 0.9 s in a temporal generalization task activated a set a cerebral areas among which the right prefrontal cortex, the anterior cingulate cortex, the right inferior parietal lobule and the cerebellum (1). The aim of the present study was to investigate whether similar cerebral areas were activated in another procedure of duration estimation, i.e, a synchronization task using an interstimulus interval longer than one second.

### Methods

Twelve normal subjects underwent 6 serial 120 second rCBF measurements with a CTI 951 R 16/31 scanner, using 60 second $H_2^{15}O$ intravenous infusions. During each scan, the subjects were presented with one of three following tasks (the order of presentation was pseudo-randomized over subjects):

A. a duration task (D) in which they had to judge the duration (2.7 s) separating successive brief illuminations of a green LED (100 ms), and to press a response button synchroneously with each LED illumination.

B. a control task (C), where the subjects had to press the response button without a hurry after LED illumination. The LED was illuminated at intervals ranging from 2 to 3.4 s, in random order (2.7 s on the average, with 0.2 s steps).

C. a force task (F), where subjects had to alternate strong and weak button presses in response to the LED illumination.

Images were analysed by using statistical parametric mapping as described by Friston and collaborators (2, 3, 4). The significance was computed for each voxel using the general linear model, at p < 0.001. The statistical inferences were calculated on SPM{Z} in terms of peak height over the volume analyzed with a threshold at p < 0.05. One comparison is reported here: the increases during the D task as compared to the C task.

Results
A significant activation was observed in the right prefrontal area (around Brodmann area – BA – 44), the anterior cingular cortices, the right inferior parietal lobule (BA 40) and vermis.

Conclusions
This experiment suggests the transprocedural invariance of the neural networks involved in the estimation of durations within a few seconds range. Most of these networks are currently attributed attentional functions (5).

1. Maquet et al. (1996) Neuroimage 3, 119–126
2. Friston K. J. et al. (1989) JCBFM 9, 690-5
3. Friston K. J. et al. (1991) JCAT 19, 634-9
4. Friston K. J. et al. (1991) JCBFM 11, 690-9
5. Posner and Petersen, (1990) ARN 13, 25–42

## T/6

THE EFFECT OF THE ANTIEPILEPTIC DRUG LOSIGAMONE ON TEMPORAL PERFORMANCE AND ATTENTIONAL COMPETENCE: *U. Ravens-Sieberer, A. Cieza, G. Leifert, B. Beyer, D. Habs, A. Dienel, N. v. Steinbüchel, E. Pöppel: Institute of Medical Psychology, University of Munich, Goethestraße 31, 80336 Munich, Germany: Sponsor: Dr. Willmar Schwabe Arzneimittel, Germany*

In a study evaluating the influence of a new antiepileptic drug (Losigamone) on mental competence in a monocentric, randomized, double-blind, placebo-controlled, parallel-group trial elenmentary, logistical and complex neurpsychological functions were measured. Since improvement of mental competence has been reported subjectively by patients taking Losigamone, one hypothesis was to assess the notion objectively whether Losigamone enhances neuropsychological functioning. As elementary logistical functions the following timing functions were measured: auditory order threshold, auditory and visual simple reaction time, delayed reactions, auditory and visual choice reaction time, finger tapping, temporal reproduction, sensory-motor synchronization, time estimation, and increment threshold for visual stimuli. Two aspects of timing can be differentiated here: central and motor aspects. As complex functions digit span, semantic memory and sequential memory were assessed.

As a first approach to assess this question an explorative study with healthy volunteers was designed.

60 healthy male, right-handed students between 18 and 30 years of age were treated with Losigamone (n = 30; 250 mg t.i.d.) or with placebo (n = 30; 250 mg t.i.d.) within a treatment period of seven days.

On the basis of a model on cognitive functioning (v. Steinbüchel & Pöppel, 1991), the results were classified into "central timing", and "motor timing" and interpreted descriptively. With respect to central timing on a meta-theoretical level the treatment group shows a minute slowing down. Such an effect is not seen for "motor timing". Central aspects of temporal integration appear to be improved in the treatment group suggesting a wider temporal platform for integrating events. In fast tapping, the treatment group is significant faster than the placebo group (in particular with right hand performance). Visual processing is improved in the treatment group, i.e. subjects under the effect of Losigamone show higher visual sensitivity presumably reflecting improved attentional competence.

These findings show that Losigamon influences cognitive functions of healthy volunteers in the above described manner, which has not been observed with other antiepileptic drugs yet. Such an influence on cognitive functions promises a very positive and hopeful effect on patients treated with Losigamone.

Steinbüchel, N. v., Pöppel, E. (1991). Assessment of mental functions in patients with epilepsy: Cognitive models and ecological Constraints. In: The Assessment of Cognitive function in Epilepsy. Edwin Dodson, W., Kinsbourne, M., Hiltbrunner, B. Demos Publications, New York 97–107

## T/7

ORDER-THRESHOLD MEASUREMENTS: RANDOMIZING VS. ADAPTIVE PROCEDURE. *A. Steffen. Institute of Medical Psychology, LMU-Munich*

Introduction: During the last years, a new perspective on temporal processing has been established: it became useful to distinguish between functions related to content (stimulus perception, stimulus processing and action such as language) and logistic functions (formal constraints, such as activation or temporal organisation). Concerning the logistic function of temporal integration, as a necessary logistical basis of the content-related functions of stimulus processing, neuronal oscillations have recently been suggested as a model for the binding of information. v. Steinb©chel (1996): Stimulus processing appears to be temporally segmented by sequential processing units of 20 to 30 ms. Such system states seem to be implemented by neuronal oscillations. These oscillations can be represented by order-thresholds. In these syszem-states information is treated as simultaneous, no order can be detected. Especially for clinical research in patients with brain lesions, a short and efficient way of measuring is needed. Contributing to this, this study is asking whether a randomized measurement of order-thresholds can be more efficient than the adaptive procedures.

Material: The order-threshold is assessed with binaural presentation of click-stimuli of 1 ms length. The subjects were asked to point to the side of the first click. The shortest interval in which the succession of stimuli could be reported right was defined as order-threshold. Two different strategies in measurement were compared: a randomized presentation of different interstimulus-intervals (ISI) varying from 80 to 15 ms and an adaptive procedure that halved the ISI in case of a right answer and added one third of the preceeding ISI in case of a wrong answer.

Method: 20 subjects were tested with both procedures in two groups with a cross-over design. All subjects were healthy volunteers and normal hearing. The average age was 30,5 years. All subjects should not have had an acoustical order-threshold measurement in the preceeding year and were hearing normal.

Results: The data suggest that both methods seem not to produce valid results of order-threshold measurement. The data were inconsistent and analyses of each step of the measurement showed that the procedures did not allow a reproduction of the measurements.

Conclusions: A short way of measuring order-thresholds is not possible by randomized testing or by a simple adaptive method. Acceptable methods have to eliminate high guessing-rates and provide reproducable testings.

STEINBÜCHEL, N. v. & WITTMANN, M. (1996): Elementare zeitliche Informationsverarbeitung als Dignoseinstrument zentralnerv-ser St-rungen. In: E. Kasten, M.R. Kreutz, B.A. Sabel (Ed.). Neuropsychologie in Forschung und Praxis. Yearbook of Medical Psychology 12, S. 146–163

## T/8

TEMPORAL SEQUENCING IN SPEECH PERCEPTION. *A. Steffen*, *A.Werani and G. Kegel. *@emsp14;Institute of Medical Psychology, LMU-Munich. Institute of Phonetics and Speech-Communication, Dpt. of Psycholinguistics, LMU-Munich*

Introduction: During the last years, a new perspective on time processing has been established. Stimulus processing appears to be temporally segmented by sequential processing units of 20 to 30 ms. Such time quanta apparently reflect system states implemented by neuronal oscillations. Contributing to this, the present study is asking whether chronological order of acoustical information in a segment up to 20@emsp14;msec is relevant or not for speech perception.

Material: The stimuli consisted of two meaningless sentences (like: It greens so grey the morning-moon) and three non-word sentence-like utterances. These stimuli had been cut up into segments, and the chronological order of the acoustical information in these segments was manipulated, i.e. the signal of each segment had been inverted whereas each segments position never changed. Each stimulus was manipulated in the segment sizes: 20, 30, 40, 50, 60 and 70 ms.

Method: The stimuli were presented via tape-recorder in different versions of manipulation. Every presentation began with the version with the biggest segments of 70 ms, descending to the version with 20 ms-segments. The unmanipulated version was presented as a control stimulus as well. After the presention of each version the subjects (48 healthy volunteers) had to transcribe their auditory impression in usual letters.

Results: Concerning to the non-word utterances, the following two important effects could be observed:

1) the shorter the manipulated speech sound (e.g. stop-consonants) in a segment, the greater the likelyhood of wrong identification.

2) the non-manipulated material could not be identified without mistakes as well.

The second effect seems to show that even if subjects try to transcibe acoustical information as it was, they tend to prefer lexical information which is found in the lexicon. No words were presented, but sometimes words were perceived. So the answers didn t even fit completely for non-manipulated material.

The first effect suggests that speech sounds temporal information within a segment of 20 ms probably is not relevant for the decoding. The stop-consonants were the shortest phonetic units and they were identyfied with the highest mistake-rate. Their phonetic information was impaired by the manipulation procedure.

An alternative method is introduced, which avoids phonetical effects in the stimuli by presenting single inverted speech-sounds in a semantic-free context. I.e. a standardized carrier-sentence like I said wug (weg, wog, a.s.o.). In this cases only one sound of the utterances last item (which always is the same non-word) should be inverted. As a control, the unmanipulated signal should be presented in a seperate session as well. The auditory impressions can be compared.

## T/9
ASSOCIATION OF TEMPORAL PROCESSING AND LANGUAGE SKILLS IN ADULTS AND CHILDREN WITH LANGUAGE DYSFUNCTIONS. *Nicole von Steinbüchel, Nikola Landauer, Marc Wittmann. Institute of Medical Psychology, Ludwig-Maximilians-University Munich, Goethestr. 31, 80336 München, Germany*

Neuropsychological and psychophysical evidence supports the notion of a strong association of language skills and general temporal processing capacities. With the experimental paradigm of auditory temporal order threshold (OT) temporal processing mechanisms on a high frequency level can be measured (OT=the minimal time interval to indicate the correct temporal order between two acoustic stimuli). OT for young healthy adults lies between 30 to 40 ms, in elderly people around 50 to 70 ms. Assessing OT in five different patient groups with acquired focal brain lesions and a patient group without brain injuries (orthopedic control group) the following results are observed: Only patients with left hemispheric posterior lesions with nonfluent aphasia showed a significantly prolonged mean OT of approx. 120 ms. Patients with left anterior lesions with aphasia (Broca s aphasia), patients with left hemispheric lesions without aphasia and patients with right anterior or posterior injuries were not significantly prolonged. This adds evidence to findings of a strong functional relationship between temporal discrimination and receptive language functions located in the temporal lobe. In previous papers (1985; 1995) v. Steinbüchel showed that the highly prolonged OT in aphasic patients can be reduced by functional training to the level of healthy subjects. Not only auditory temporal discrimination but also discrimination of phonemes are significantly ameliorated. These findings of an association of time and language functions and their successful improvement by training have recently also been found in language-learning impaired children (Merzenich et al., 1996; Tallal et al., 1996). Furthermore we report data on auditory OT and phoneme discrimination in 110 children of the second year of elementary school and of 24 children with reading and writing impairments, taking part in special training classes. OTs were significantly lower in normal pupils than in the children with reading and writing impairments. In both children groups the ability of phoneme discrimination was associated with the performance in OT.

The ability to discriminate between the phonemes /da/ and /ta/ was significantly better in children with lower OTs than in children with higher Ots. First training trials of auditory OT in children with reading and writing impairments show promising results (Supported by BMBF).

## T/10
REPRODUCTION OF TEMPORAL INTERVALS: RELATION TO NORMAL COGNITIVE DEVELOPMENT. *E. Szelag[1], K. Rymarczyk[1], J. Dreszer[1], N. v. Steinbüchel[2], E. Pöppel[2]. [1] Nencki Institute of Experimental Biology, Department of Neurophysiology, 02-093 Warsaw, 3 Pasteur Street, POLAND; [2] Institute for Medical Psychology, D-80336 Munich, Goethestrasse 31, GERMANY*

A growing body of evidence on temporal mechanisms in human information processing indicates that successive, elementary events can be linked together by an integration mechanism. The experimental support for such temporal integration (TI) with an upper limit of a few sec. comes, for example, from studies using a paradigm of nonverbal reproduction of temporal intervals. Accordingly, intervals of app. 3 sec. are reproduced veridically. Shorter intervals are usually slightly overestimated, whereas those longer ones are reproduced shorter than their real duration. The indifference time interval between over- and underestimation has been interpreted as the temporal limit of TI. Two experiments reported here tested whether these relationships may be altered during normal cognitive development. A further purpose was to investigate whether the modality (visual vs. auditory) or the temporal domain of presented stimuli can influence temporal limits of this hypothetical mechanism. Thirty eight right-handed school pupils aged from 9–10 (younger subjects) or from 13–14 years (older subjects) were studied. A green light or a pure tone of a frequency of 200 Hz were used as stimuli. The exposure time of a target stimulus varied from 1 to 5.5 sec (Experiment 1) or from 1 to 3 sec (Experiment 2) in steps of 0.5 sec. The subject's task was to reproduce the duration of presented target by pressing a button. The main finding was the absence of developmental differences in temporal limits of TI. Intervals of app. 2.5 sec were reproduced appropriately, independently of both the stimulus modality and the temporal domain of presented stimuli (the target duration from 1 to 5.5 or from 1 to 3 sec). Moreover, the results indicated that shorter intervals were overestimated and longer ones underestimated. It can be concluded that TI reflects a fundamental supramodal process with a consistent time limit from 9 to 14 years. Thus, the age-related increases in cognitive abilities in children across ages 9 to 14 years do not affect TI.

## T/11
THE VISUO-SPATIAL COMPONENT IN CLOCK READING – DC POTENTIALS RELATED TO ANALOGUE VERSUS DIGITAL CLOCKFACES. *F. Uhl[1,3], Ebenbichler[2], W. Lang[1], L. Deecke[1], G. Kerkhoff[3]. [1] Neurologische Universitätsklinik, Währinger Gürtel 18, A-1090 Vienna, [2] Institut für Physikalische Medizin, Währinger Gürtel 18, A-1090 Vienna, [3] Entwicklungsgruppe Klin. Neuropsychologie, Dachauerstr. 164, D-80992 Munich-Bogenhausen*

The study attempts to localize the cortical substrate of processing information on time being either presented by an analogue or a digital clock face. Event-related DC potentials were recorded time-locked to the tachistoscopic presentation of three clockface slides, conveying either analogue or digital information on time. Two seconds after the first one, the second slide was presented. 8 sec after that, the third was presented.

here were four conditions: In the two arithmetic tasks (one of those was digital and one was analogue), subjects had to calculate and judge whether or not the third clockface conveyed the exact difference between the time showed by the first and the second slide. In the two control conditions, subjects only attended to (analogue or digital) time displays without any requirement to perform manipulations on the information conveyed.

In the analogue conditions, right hemispheric lateralisation of negative DC potential shifts was observed at frontolateral and at anterior temporal sites. At the latter sites, lateralization was present only when calculations had to be performed (i.e. in the *arithmetic* analogue condition). Furthermore, these calculations on analogue clockfaces resulted in an decrease of negative shifts at anterior sites and an increase at all retrorolandic sites, possibly due to vivid mental imagery strategies that have been shown to lead to high retrorolandic DC shifts [Uhl, Goldenberg et al. Neuropsychologia 28:81–93, 1990].

By contrast, there was no hemispheric lateralisation with digital clockfaces. The report is going to appear in Neuropsychologia [1997, Vol. 35].

## T/12

EFFECT OF ACOUSTICAL STIMULATION AND EMOTIONAL STATE ON TIME ESTIMATION. *A. Venneri & E. Gilhooly. Department of Psychology, University of Aberdeen, UK*

Evidence from past research shows that adults' understanding of time is neither natural nor intuitive, but is the result of a gradual, constructive process. Psychological time is an abstract, uniform, measurable dimension that extends without limit into the past and the future. Views of time vary from one culture to another. This diversity may imply differences in the way people conceive and experience time.

There is also evidence to suggest that certain individual differences produce differences in attitude towards time. It has been shown that differences in personality produce differences in temporal experience and in people's behaviour. For example: people who are competitive, hard-driving, or angry show a sense of time urgency that causes them to overestimate the time spent on a task. However, subjective time experience may be even more strongly affected by external factors that may alter either the physiological or the psychoemotional balance of individuals. For example, drug consumption alters an individual's experience of time considerably. Of special interest and clinical importance are the effects of mental illness (depression, schizophrenia) upon time estimation, although findings on this topic are sparse and often confusing.

Very little is known of the effects upon perceived duration of acoustic variables, or of emotional state, either in healthy subjects or in patients with mental illness. One study with young healthy subjects found that they tended to underestimate clock time when listening to a piece of music.

The purpose of the present study was to explore the effects upon time estimation of emotional states and acoustic loading. Subjects were selected among undergraduate psychology students using the Zuckerman and Lubin 'Multiple affect adjective checklist'. Only subjects scoring at the two extremes of the distribution for depressive mood were included. The experimental sample included 18 subjects with no tracts of depression and 12 subjects who achieved the maximum score in the depression section of the checklist. All subjects performed a verbal time estimation task (from 10 to 40 seconds) in three conditions (silence, listening to a piece of classical music, listening to the noise of a drill). Silent counting was prevented by asking the subjects to read numbers aloud.

Results showed a significative effect of the acoustic stimulation (p < 0.01) and a significative interaction between group and acoustic stimulation (p < 0.02): subjects without depressive tracts underestimated time while listening to the music or the noise; on the contrary, subjects with depressive tracts showed no difference between the silence condition and the music condition, but significantly overestimated the intervals under the noise condition.

These results suggest that the physical characteristics of the acoustic loading affect time estimation, and these characteristics interact with an individual's initial emotional state.

Address correspondence to: Dr. Annalena Venneri, Department of Psychology, University of Aberdeen, King's College, ABERDEEN, AB24 2UB, UK, Tel. +44-1224-273483, Fax. +44-1224-273426, e-mail: annalena@abdn.ac.uk

## T/13

LEFT HEMISPHERIC DOMINANCE AND FUNCTIONAL DISSOCIATION FOR TEMPORAL ORGANISATION OF AUDITORY PERCEPTION AND PERSONAL TAPPING TEMPO. *M. Wittmann, N. von Steinbüchel. Institute of Medical Psychology, Ludwig-Maximilians-University, Goethestr. 31, 80336 München, Germany*

Evidence for a strong association of language competence and auditory temporal resolution has been collected for over three decades. Patients with aphasia and children with language-learning-impairments often show deficits in the ability to indicate the temporal order of two consecutive stimuli, as measured i.e. with a temporal order discrimination task. In our study we first wanted to replicate these findings and increase knowledge through the measurement of temporal order thresholds. As there are findings of a commen processing mechanism for perception and action secondly the question was investigated if patients with aphasia also show an impairment in temporal aspects of simple motor tasks as measured with the personal and the maximum finger tapping speed; in both tasks the Interresponse-Intervals (IRI) were assessed.

Five patient groups with acquired focal brain injuries and an orthopedic control group were investigated: Patients had brain lesions of the 1. left hemisphere anterior (with nonfluent aphasia; LH anterior), 2. left hemisphere posterior (with fluent aphasia; LH posterior), 3. left hemisphere with predominately subcortical lesions (LH without aphasia), 4. right hemisphere anterior (RH anterior), 5. right hemisphere posterior (RH posterior). These groups were selected to test the hypothesis of a special localisation of temporal aspects as measured with the three experiments mentioned above. Prior to that we tested the hypothesis of a general lateralisation of functions in forming a group with lesions of the left hemisphere and aphasia (LH: group 1 and 2) and a group with lesions of the right hemisphere (RH: group 4 and 5).

Results for the general lateralisation hypothesis show that only the group LH has a statistically significant prolonged mean order threshold (a twofold increase as compared to the control group). In the personal tapping speed only the group LH shows a statistically significant prolonged mean IRI (a twofold increase). In the maximum tapping task neither the group LH nor the group RH is slowed down. Results for the special localisation hypothesis indicate that only the group LH posterior has significantly prolonged order thresholds. Only the group LH anterior shows to be significantly slowed down in the personal tapping tempo. In the maximum tapping tempo there are no group differences. All brain injured patient groups are able to perform finger movements with maximum speed.

Additional correlational and single case examinations confirm that both the ability of auditory temporal order discrimination and the temporal organisation of voluntary movement as measured with the personal tapping task are associated mainly with the left hemisphere. A dissociation of the underlying functions can be registered intrahemispherically. Distinct neural algorithms of the left hemisphere for the two temporal processing mechanism seem to be activated. The results imply also that the performance of a personally, self paced tempo and a maximum tempo are processed differentially (Supported by BMBF).

# Vision

## V/1

INTERHEMISPHERIC TRANSFER OF VISUAL MOTION IN-
FORMATION IN A CASE OF POSTERIOR CALLOSAL LESI-
ON: NEUROPSYCHOLOGICAL AND fMRI EVIDENCE. *S. Clar-
ke[1], P. Maeder[2], R. Meuli[2], F. Staub[1], A. Bellmann[1], L. Regli[3], N. de
Tribolet, G. Assal[1]. [1] Division de Neuropsychologie, CHUV, 1011
Lausanne, Switzerland, [2] Service de Radiodiagnostic, CHUV, 1011
Lausanne, Switzerland, [3] Service de Neurochirurgie, CHUV, 1011
Lausanne, Switzerland*

We report the case of a 23 year old right-handed man, who suffered
a calloso-occipital haemorrhage from a arterio-veinous malformati-
on of the splenium of the corpus callosum. Six months after the haemorrhage, a MRI showed complete destruction of the posterior half
of the corpus callosum and hypodense lesion in left precuneus. On
tachistoscopical examination, the patient was unable or greatly impaired to read letters, words or geographical names or to name pictures or colours presented within left hemifield. Interhemispheric
transfer of motion information was tested by presenting short sequences of coherent dot motion within left hemifield while the patient was fixating a central point. Eye movements were recorded
with a camera and trials with saccades were rejected off-line. The
patient reported accurately on coherent dot motion presented in
the left hemifield.
Activation pattern to apparent motion stimuli were studied with
fMRI in this patient and in 10 age-matched normal subjects. The stimulus was a red-black checkerboard of $6 \times 6$ squares (changing at
8 Hz) with a central fixation point. The control situation was rest
in darkness, with eyes open. Three conditions were studied: i) bilateral stimulation ; ii) unilateral left (left half of checkerboard); and
iii) unilateral right stimulation. Measurements were conducted on
a 1.5T system with a surface coil on the occiput. In each subject,
7 coronal images of the occipital and posterior parietal lobes were
acquired during 6 on/off epochs. FMI maps were generated by identifying voxels correlated with stimulus time course (Z score = 0.7).
Regions of interest were defined in the calcarine region (V1) and
on lateral occipital gyri (including V5/MT). In all normal subjects,
apparent motion stimuli activated strongly striate and extrastriate
cortex. When apparent motion stimuli were presented to one hemifield only, the contralateral calcarine region was activated similarly
as in the bilateral condition; the ipsilateral calcarine region was either not activated or to a much lesser degree. Regions on the occipital convexity were activated similarly on the contra- and ipsilateral
side. The patient presented a similar activation pattern; in particular,
unilateral stimulation was accompanied with activation on contralateral and ipsilateral occipital convexity.
In conclusion, our patient with posterior callosal disconnection
showed preserved interhemispheric transfer of motion information
both in behavioural and fMRI studies.

## V/2

IMPAIRED DELAYED AND ANTI-SACCADES IN A VISUAL
FORM AGNOSIC. *H.C. Dijkerman, A.D. Milner, & D.P. Carey*.
School of Psychology, University of St. Andrews, Fife, KY16 9JU,
UK. *Neuropsychology Research Group, Department of Psycho-
logy, King's College, University of Aberdeen, AB9 2UB, UK*

Milner & Goodale (1995) proposed that the ventral and dorsal streams of visual cortical processing mediate visual perceptual and visuomotor functions respectively. The visuomotor functions of the
dorsal stream require continuously updated visual information,
coded in an egocentric frame of reference. The visual perception
mediated by the ventral stream requires allocentric, viewpoint-inde-
pendent coding of visual information which can be stored over long
periods of time. The current study assessed performance of a visual
form agnosic, DF, whose brain lesion mainly affects the ventral
stream, on three versions of a saccadic eye movement task. In the
control task, the subjects made a saccade to a target as soon as it became visible. In the delayed saccade task, the subject was instructed

to wait 5 sec. before making the saccade to the position where the
target had appeared. The anti-saccade task required the subject to
saccade to the mirror position of the target. It was hypothesized that
the latter two tasks would involve the ventral stream of visual processing. Compared with normal control subjects, DF's saccades
were less accurate in the delayed and anti-saccade tasks, but not
in the control task. These findings suggest a role for the ventral
stream in 'off-line' spatial visual processing.
Milner AD & Goodale MA. The Visual Brain in Action, 1995, Oxford University Press.

## V/3

TIME-TO-ARRIVAL ESTIMATES FOR VISUAL TARGETS
MOVING UPWARD OR DOWNWARD WITH CONSTANT
SPEED: *W.H. Ehrenstein: Institut für Arbeitsphysiologie, Abt. Sin-
nes- und Neurophysiologie, Ardeystr. 67, D-44139 Dortmund, Germany*

When an object moves, it provides not only sensory information regarding its present speed, but also information about its future position and arrival time. *Time-to-arrival information* is vital for any
temporally tuned perceptual-motor act such as catching a ball or
crossing a busy street. Time-to-arrival studies have, to date, almost
exclusively addressed motion in the sagital plane and have concentrated on the optic variable $t$ (i.e., the rate of expansion of an approa-
ching object on the retina). Time-to-arrival estimates in the absence
of expansion cues, namely, for *vertical* motion in the *frontoparallel*
plane have received relatively little attention. In this case, objects typically accelerate or decelerate due to *gravity* so that the vertical
component of acceleration (e.g., as in the rise and fall of a thrown
ball) is roughly constant at $9.8 \text{ m s}^{-2}$. Consequently, time-to-arrival
estimates might be affected by supposed gravitational acceleration
even in cases where upward and downward motion occur at a constant speed. In the present study on *extrapolated* motion, the target
was a spot of light displayed by a vector-scan on a fast-phosphor
(P-31) screen. It moved (starting 12° above or below fixation) upward or downward at $6° \text{ s}^{-1}$ along the vertical meridian for 2 s, so
that it vanished at the position of fixation. Subjects indicated by
key press the moment at which the target would reach a stationary
reference if it continued to move as displayed. The reference was located randomly at one of five positions, each 2° apart, so that the
distance over which stimulus motion was extrapolated ranged between 2° and 10°. Data from four subjects, tested in six blocks of
30 trials, showed a dependence on motion direction. Averaged
across blocks of trials, extrapolation distances, and subjects, time-
to-arrival estimates were *shorter* by 62 ms for *downward* than for
upward motion (p < 0.04). This difference depended markedly on
the amount of practice with the task. It was largest (91 ms) in the
first block of trials, but it decreased with the number of blocks to
14 ms in the sixth block. In addition, the difference in time-to-arrival
estimates between upward and downward motion increased with extrapolation distance (from 57 to 202 ms) in the first block, but no
such dependence on extrapolation distance was found in the final
block. In sum, the data indicate that time-to-arrival estimates seem
to be influenced (or biased) to some extent by previous experience
with accelerating or decelerating effects of gravity, even when stimulus speed is constant, in that they tend to be shorter for downward
than for upward motion. This bias, however, is strongest in initial
trials and diminishes rapidly with repeated and exclusive exposure
to moving stimuli of constant speed, suggesting that time-to-arrival
estimates may *adapt* according to the prevailing stimulus acceleration.

**V/4**

RAPID ASSESSMENT OF THE CONTRAST SENSITIVITY FUNCTION IN THE HOODED RAT. *J. Keller[1], H. Strasburger[1], D. T. Cerutti[2] & B. A. Sabel[1]. [1] Institut für Medizinische Psychologie, Otto-von-Guericke-Universität Magdeburg, Leipziger Str. 44, D-39120 Magdeburg, Germany, [2] Department of Psychology, Davidson College, Davidson, NC 28036, USA*

The spatial contrast-sensitivity function (CSF) is a sensitive, quantitative test for assessing spatial visual function that is comparable across species. Rat vision is widely assumed to be too poor for meaningful psychophysical assessments, and the perceptual characteristics of vision in rats, despite their predominate laboratory use, have been rarely studied. This neglect is due partially to the difficulty in behaviorally determining the CSF efficiently and accurately. We have developed a method for rapid assessment of the CSF in the hooded rat using a computer monitor for stimulus display and an infrared touch screen as response input device. Sine-wave gratings of variable contrast and spatial frequency were presented in a 6-alternative forced-choice task; a rat's nose-poke to the target stimulus produced reinforcement (water) and nose-pokes to other locations repeated the trial with a short, aversive time-out. Spatial frequencies assessed were in the range of 0.036 to 0.39 cyc/deg; at each spatial frequency tested, stimulus contrast was varied according to a simple adaptive procedure. Psychometric functions were determined by fitting a Logistic to the binary response data through a maximum-likelihood fitting procedure (Harvey, 1997), and the point of inflection was taken as the threshold. The functions obtained had the typical inverse-U shape, with peak sensitivity occurring at around 0.10 cyc/deg, similar to previous data. We find this procedure to be a valid method for rapidly determining the rat's CSF, presenting us with a tool for assessing spatial vision after experimental manipulations of the visual system.

**V/5**

WHAT IS LEARNED IN VISUAL SEARCH – LOCAL BRIGHTNESS CONTRAST OR UNIQUE VISUAL ATTRIBUTES? *U Leonards, R Rettenbach, G Nase, R Sireteanu. Department of Neurophysiology, Max Planck Institute for Brain Research, Deutschordenstr. 46, 60528 Frankfurt, Germany*

Serial visual search can become parallel with practice (Sireteanu & Rettenbach, 1995, *Vision Research* 35 2037–2043). We wondered whether this improvement in performance depends on local brightness differences between target and distractors or on different visual attributes. In a first experiment, we compared the dynamics of learning of a task in which the target had a unique visual attribute but there were no brightness differences (lines of different orientation: "tilt") with that of a task defined by a visual attribute and a local brightness difference (circles with an added vertical line among plain circles: "added line"), and a task in which the differences between target and distractors consisted only of a brightness difference (circles with a gap of 45° among circles with a gap of 135°: "gap size"). In a second experiment, we tested whether learning based only on local brightness differences ("gap size") transferred to a task in which a visual attribute and a local brightness difference were present (a circle with a gap of 90° among plain circles: "gap"). In a third experiment, subjects were tested in a task in which neither a local brightness contrast nor a unique visual attribute were present but rather the target was defined by a conjunction of attributes ("conjunction" of colour and orientation). The subjects' task was to indicate the presence or absence of a target on a computer screen by immediately pressing a button and pointing to the location of the target if the trial was positive, or raise the hand if negative. No feedback was given. Response time and error rate were recorded. Eight naive and two experienced subjects participated in at least 14 experimental sessions. We found that both brightness differences and visual attributes lead to parallelisation with practice of originally serial search. However, transfer experiments indicate that there is no transfer but rather trade-off between the learning induced by brightness or visual attributes. Tasks devoid of local brightness and of unique visual attributes ("conjunctions") did not become parallel even after prolonged practice.

**V/6**

EVIDENCE AGAINST BLINDSIGHT IN NORMAL OBSERVERS. *L. Muckli, W. Singer and R. Goebel: Max-Planck-Institut für Hirnforschung, Deutschordenstrasse 46, 60528 Frankfurt/Main*

Blindsight (1) is the ability of patients with damage to the visual cortex to discriminate visual stimuli without conscious perception. Kolb and Braun (2) described two experimental paradigms that induce blindsight phenomena in normal subjects. One of the two is based on opponent moving dots that are usually perceived as transparent surfaces of random dot patterns. If, however, opponent moving dots are paired and presented in close spatial proximity, motion is no longer perceived, and the display is perceived as a set of incoherently flickering dots. Kolb and Braun reported that in these conditions, targets of orthogonal trajectories, shortly presented in one quadrant of the display, were detected as well as in transparent motion control conditions but showed no correlation to subjective confidence ratings. We reproduced Kolb and Braun's procedure of transparent motion stimulation with only minor changes but observed no blindsight effects.

Two parameters, the dot distance within the pair of opponent moving dots and the dot lifetime, were varied in 25 stimulus constellations. Six observers were asked to report the target position (1 of 4 possible) and to rate their confidence (scale of 1 to 4) after viewing each trial for 250 ms.

Solid symbols in **a** represent the averaged performance in all 25 stimulus conditions, open symbols represent the averaged confidence ratings. The correlation of success and confidence is shown in **b** for four individuals averaged over all conditions. Data were analysed individually and whenever performance was well above chance (25%) confidence ratings were correlated to success. Clear transparent motion was only perceived if the distance moved by dots exceeded the distance within a pair. At small distances dot pairs are perceived as one smeared line, hence orthogonal targets were detected easily. In close-to-threshold conditions subjects were usually strained by the task. Some subjects reported to perceive target areas as being brighter or more dense. These verbal reports are believed to be post hoc rationalisations for the segmentation of the image without a clear motion orientation cue. Training is known to increase the ability of blindsight. Nevertheless two subjects showed no increase of performance after a week of daily exercise.

Kolb and Braun's subjects were instructed to use the full confidence scale, irrespective of their absolute sense of certainty, which might have tempted subjects to falsely use high-confidence judgement (opposite of blindsight). Display parameters of the Kolb and Braun "paired"-stimulus was systematically varied from "blind" conditions in which no detection was observed over conditions of slightly increased performance to good performance. We could not find any conditions in which detection and confidence were clearly dissociated as required for blindsight.

This notion is further supported by a recent report of Morgan et al. (3) who analysed the second paradigm of the Kolb and Braun study (not tested here). Again the stimulus paradigm was insufficient to induce blindsight in normal observers.

1. Pöppel, E., Held & R. Frost, D. Nature 243, 295–296 (1973)
2. Kolb, FC. & Braun, J. Nature 377, 336–338 (1995)
3. Morgan, M. J., Mason A. J. S. & Solomon J. A. Nature 385, 401–402 (1997)

## V/7

ELECTROPHYSIOLOGICAL EVIDENCE OF A PERCEPTUAL PRECEDENCE OF GLOBAL VS. LOCAL INFORMATION. *A. M. Proverbio\*, A. Zani & A. Minniti. Department of Psychology, University of Trieste, Via dell'Universita' 7, 34123 Trieste, Italy; \*Istituto di Psicologia, National Research Council, Viale Marx 15, 00137 Rome, Italy*

The goal of the present study was to investigate visual processes involved in global vs. local analysis of visual configurations by means of event-related potential recording (ERPs) and reaction time measures (RTs). A *global precedence* effect has been widely described in literature as the evidence that global structure interferes with local pattern processing more than local structure interferes with global pattern processing. Recently Hughes et al. (1996) provided strong psychophysical evidence that global precedence might depend on a perceptual advantage of low spatial frequency processing in the visul system. To test this hypothesis we recorded ERPs and RTs from 8 right-handed young volunteers using hierarchical letter stimuli in a selective attention task. Stimuli were large S or H letters composed of smaller S or H letters. Due to the combination of letters at the two levels stimuli might be congruent, when they had the same identity (i.e., S/S or H/H), or incongruent, when they had different identity at the two levels. (i.e., S/H and H/S). Stimuli were equiprobable and randomly presented for 100 msec at two eccentric locations of the right and left visual field along the horizontal meridian. Subjects were asked to gaze the fixation point, and to pay covert attention and respond as accurately and fast as possible to target letters at either the local or global level with their right or left hand.

In agreement with current literature, RT data showed significantly faster responses to global than local letters. Furthermore, RTs to local targets were significantly slowened when the global structure was incongruent, and not vice versa, thus showing an interference effect of the global toward the local level. Very interestingly, ERP amplitude measures of N115 sensory component provided evidence of a perceptual precedence of global vs. local processing at primary visual areas. In fact, early visual evoked responses were significantly smaller when the target was incongruent at the local level with the global structure. Conversely, sensory responses were not affected by local information when the target was at the global level. These data strongly support the hypothesis of a *perceptual precedence* of global information in terms of a dominance (or facilitation) of visual outputs conveyed by *low frequency channels*.

## V/8

INFANTILE STRABISMUS. *S. Réthy. ophthalmological practice, Kreuzstraße 39, D-46535 Dinslaken*

DONDERS' THEORY 100% VALIDATED
correcting congenital errors with early glasses
Method: Initially intermittent effort/relaxation has to be differentiated from rapidly provoked adaptations against relaxation with glasses given late, after adaptive resistance was programmed by delay in the brain of infant.
The effort of near reaction (accommodation/convergence) applied for far distance will be latent, not detectable with glasses anymore. Temporary manifestation needs repeated cycloplegia. (DONDERS)
Results: The early diagnosis has shown the accommodative origin of initially intermittent effort. It was relaxed with early glasses in 146 cases (during the last 3 years), before the rapid adaptations in the brain had been provoked: at 2nd – 3rd month of age.
Congenital important HYPEROPIA means constant blur stimulating accommodation effort for far distance. At 2nd – 3rd month of age +4.00 or more hypermetropia detected and corrected immediately removes the blur stimulating the effort for far distance as well. At 4th month of age adaptations of brain are provoked, it is too late for correction of pure accommodative/convergence effort getting latent.
Conclusion: Diagnosis has to detect
a) the blur (stimulating the effort) weakening the fusional check of convergence applied for far in the eye

b) the intermittent esodeviation (provoking suppression, shift of localization, tonus stabilization) in the brain.
Surgery cannot prevent, when suddenly achieving straight position of eyes in a late stadium. For the brain program adapted to esodeviation it is a shock.
Diagnosis of full accommodative origin has to precede the full clinical picture of infantile esotropia containing adaptations provoked in the brain irreversibly establishing monofixation. It is an addition to the pure accommodative causes remaining latent, hidden under the adaptations due to the delay at 2nd – 3rd month of most sensitive age.

## V/9

THE FAST-BRAIN IN ACTION VS THE SLOW-BRAIN IN IDENTIFICATION. *Yves Rossetti, Laure Pisella, Gilles Rode, Marie-Thérèse Perenin, Céline Régnier, Mohamad Arzi, Dominique Boisson. INSERM, U 94: Vision et Motricité – 16 avenue Lépine, 69500 Bron – France. – E-mail: rossetti@lyon151.inserm.fr*

There is a convergence of data from various field to support the dissociation of visual pathways into two main streams projecting from occipital to frontal cortex via the posterior parietal lobe (dorsal route) and via the inferotemporal lobe (ventral route). It is usually assumed that the dorsal route drives information that is usefull for driving an action toward the stimulus (i.e. metric properties; e.g. localisation), whereas the ventral route extracts information usefull for identifying it (i.e. intrinsic properties; e.g. colour). In addition, electrophysiological studies have shown that inputs to the ventral stream lag those to the dorsal stream by about 12 ms (1). Arguments for two streams of sensory processing dissociated in time will be taken from three lines of reasearch:

1) Blindsight (2) and Numbsense (4) patients have been studied for their ability to act toward a stimulus and their capacity to describe it manually or verbally. Only goal directed responses showed above-chance performance. A disruption of these abilities was observed when stimuli had to be memorised for a short delay (< about 1 s). This suggests that only a specific pathway devoted to action can be afferented in these patients.
2) The same effect was investigated in normal subjects. Pointing to memorized targets were compared to immediate pointing. Different frames of reference were observed between these two conditions. When a target had to be memorized, pointing distributions were affected by allocentric cues, which reveal a higher level processing of spatial information (3).
3) Our last study compared latencies of visuomotor processing for a ventral attribute (colour) and a dorsal attribute (location) during a pointing task. Target location and/or color was altered upon movement onset. Instructions were to correct movement direction or to interrupt the movement according to the target change. It was found in both cases that colour processing was slower (by about 80 ms ) than location processing of the same target. Performance observed for identical movement speed was always higher for location responses. This result was confirmed by the observation that movement duration spontaneously chosen by subjects was longer when they had to process color.
Taken altogether, these data suggest that among the various stream of sensory processing observed in the brain, it is possible to identify a specific system exclusively devoted to action (5).

**References:**
1. Nowak, L., & Bullier, J. (1996). The timing of information transfer in the visual system. In J. Kaas, K. Rochland, & A. Peters (Eds.), Extrastriate cortex in primates. in press
2. Perenin, M.-T., & Rossetti, Y. (1996). Neuroreport, 7(3), 793–797
3. Rossetti, Y., & Régnier, C. (1995a). Representations in action : pointing to a target with various representations. In B. G. Bardy, R. J. Bootsma, & Y. Guiard (Eds.), Studies in Perception and Action III. (pp. 233–236). Mahwah, NJ: Lawrence Erlbaum Associates, Inc
4. Rossetti, Y. et al. (1995b). Neuroreport, 6, 506–510
5. Rossetti, Y. (1997). Implicit perception in action: short lived motor representations of space. In P. Grossenbacher (Ed.), Finding consciousness in the brain. Benjamin Publishers, in press

## V/10

**THE PLACE PREFERENCE TASK, A TASK ALLOWING NEW PERSPECTIVES IN STUDYING THE RELATION BETWEEN BEHAVIOR AND PLACE CELL ACTIVITY IN RAT.** *J. Rossier[2], Yu. Kaminsky[1], F. Schenk[2] and J. Bures[1]. [1] Institute of Physiology, Academy of Sciences, 142 20 Prague 4 – Krc, Czech Republic, [2] Institute of Physiology, University of Lausanne, Rue du Bugnon 7, 1005 Lausanne, Switzerland*

A great number of researches studying place cell activity have been done, using a standard pellet-chasing task. But it could be possible that rats did not use any spatial representation to succeed in that task. Therefore, it seemed important to create a task in which rats *had to* use a spatial representation. In that goal, we have developed a setup which combined a random pellet-chasing task with a goal directed navigation task.

Testing was conducted using a circular elevated open field (∅ = 80 cm) with a unmarked target area (∅ = 20 cm) in the center of one of the four equal quadrants. The computer-controlled tracking system released a pellet, randomly available on the open field, whenever the rat had entered this target area and stayed there for an uninterrupted interval of 0.3 second. Once the reward had been delivered, the rat had to search it and to stay outside the target area for at last 3 seconds before it could make a new rewarded visit of the target. This setup allowed alternation between random pellet-chasing and goal-directed place navigation.

To evaluate which type of information the rat used to resolve this task, we have trained the animals in 5 conditions: a) under light (unlimited number of visual cues); b) in darkness with two visual cues, each of them formed of a set of 3 LEDs; c) in darkness with no cue; d) in darkness on moving arena rotating at an angular speed of one revolution per minute, with two cues; and e) in darkness on moving arena with no cue.

In these experiments, it appeared that rats were able to learn this place preference task, under light as well as in total darkness on stable or moving arena. The switch from the light condition to the darkness with two cues condition, as well as from darkness to darkness and moving arena conditions, induced a drastic decrease of performances. This pointed out that visual information played an important role under light and that rotation disrupted at first the path integration solution of the task. The deletion of both cues had, on the contrary, no impact on the performances. Under light, animals were able to switch quickly to a new target position, whereas it takes more time in darkness. Rats seemed to have a more solid representation of space under light, and path integration seemed to be an effective way to guide navigation in darkness in small areas.

This place preference task will allow us to study PCs when rats *have to* use their spatial representation. It will be then possible to compare the behavioral response with the PCs activity.

Supported by IGA AVCR Grant 711401 and by the Société Académique Vaudoise.

## V/11

**VISUAL PERCEPTION AND PERFORMANCEINTELLIGENCE AFTER CEREBRAL VISUAL IMPAIRMENT DUE TO NEONATAL BRAIN DAMAGE.** *P. Stiers[1,2], P. De Cock[2], and E. Vandenbussche[1]. [1]Laboratory of Neuropsychology, K.U.Leuven, Medical School, Herestraat 49, B-3000 Leuven, Belgium; [2]Centre for Developmental Disabilities, University Hospital, Kapucijnenvoer 33, B-3000 Leuven, Belgium*

We investigated whether impaired performance on visual perceptual (VP) and performance intelligence (PI) tasks, which are strongly associated in neonatally brain damaged children (Ito et al., 1996, *Dev Med Child Neurol*, 38: 496–502), reflect separate deficits. Therefore, the performance of 22 neonatally brain damaged children with cerebral visual impairment (CVI) and 16 mentally retarded (MR) children with no CVI-indications was studied on a visual object recognition task (the De Vos-task), in relation to intelligence data available from clinical records. First, we investigated whether deficits occurred that are specific to the VP domain. This is the case if

performance on the De Vos task is weaker than can be expected from the child's PI level. This is established by comparing the child's De Vos score to that of normal children of an age corresponding to the child's PI-age. According to the above criterion, 16 CVI children (73%) were specifically VP-impaired, compared to only three MR children (19%). Secondly, we investigated whether this VP-deficit affects performance on intelligence tests. It was found that the ratio of performance to verbal intelligence was significantly lower in VP-impaired children (14 WPPSI forms, 8 VP vs. 6 nVP: $F(1,12) = 10.7$, $p = 0.0035$). On the other hand, no larger performance subtest scatter was found in VP-impaired children (29 SON forms, 16 VP vs. 13 nVP: $F(1,27) < 1$), nor did they show a different subtest profile (group × subtest interaction: $F(4,108) < 1$). These results show that neonatal brain damage can give rise to deficits that are visual perceptual in nature, eventhough these deficits seem to go together with reduced PI. If this PI reduction was a mere performance reduction due to VP-impairment, one would expect some PI subtests to be more vulnerable, or at least an increase in subtest scatter. Since this was not found it might reflect a separate impairment.

## V/12

**THE DIFFERENTIAL CONTRIBUTION OF THE MAGNOCELLULAR AND PARVOCELLULAR CHANNELS IN PROCESSING OF VISUAL INFORMATION IN THE TWO HEMISPHERES OF THE HUMAN BRAIN.** *Iwona Szatkowska and Anna Grabowska. Nencki Institute of Experimental Biology, Department of Neurophysiology*

The processing of visual information in primates is accomplished by two parallel visual pathways: magno- and parvocellular channel. The magnocellular channel is more sensitive to low spatial frequencies. This system is thought to be involved in global analysis of visual scenes. The parvocellular channel is more sensitive to high spatial frequencies and is involved in identification of visual patterns, especially small, local details. It has been hypothesized that the two hemispheres differ in their ability to process visual information carried by these two visual channels (Sergent 1983, 1987). The present experiment aimed at testing this hypothesis by using a task in which figures of various sizes and complexity were compared. The stimuli were presented in pairs, one after another, each for 100 ms, at an interstimulus interval of 50 – 500 ms. The subject's task was to indicate (by pressing one of three buttons) whether the second stimulus was the same, smaller or bigger than the first one. The first stimulus in each pair was exposed unilaterally, randomly in the left (LVF) or right (RVF) visual field, and the second one was presented at the centre of the visual field. The reaction time analysis shoved significant interaction between stimulated hemifield and stimulus size, and between stimulated hemifield and stimulus complexity. Small and more complex stimuli were processed faster in RVF presentation conditions than in the LVF presentation conditions. Large and less complex stimuli were processed faster in LVF presentation conditions than in the RVF presentation conditions. Our data support the view that the two hemispheres may differ in their ability to process visual information carried by magno- and parvocellular channel.

## V/13

**DOES IMAGERY ACTIVATE THE PRIMARY VISUAL CORTEX ?** *A.Thiel[1], H.J. Markowitsch[2], J. Kessler[1], W.D.Heiss[1]. [1] Max-Planck-Institut für neurologische Forschung, Gleuelerstr. 50, 50931 Köln, [2] Physiologische Psychologie, Universität Bielefeld, Postfach 100131, 33501 Bielefeld*

<u>Objective:</u> It is still a matter of controversy whether the primary visual cortex is activated during visual mental imagery. Whereas recent PET studies of Roland et al [1] did not report any activation in primary visual cortex during recall of imagined colored, complex geometrical patterns, other PET and functional MRI studies [2, 3] showed different results. One major problem with present functional MRI or PET studies on visual imagery is the localization of the tar-

get area. As high individual variance in the anatomy of the calcarine fissure makes localization by means of visual inspection or stereotactic coordinates difficult. We used a functional definiton of the primary visual cortex, (areas V1 and V2) to account for individual anatomical and physiological variation. In an O-[15]-water PET study 5 normal volunteers were required to imagine abstract pictures with their eyes closed compared to a resting state. Visual activation with an alternating checkerboard was used for individual definition of the region of interest.

Methods: 4 abstract pictures were presented to 5 healthy, male volunteers, 24 to 39 years old, without any neurological or psychological disorders half an hour prior to the PET examination. The persons were instructed to learn the pictures and to make up characteric keywords for each image. The PET examination comprised 12 measurements of relative cerebral blood flow (CBF) using O-[15]-water bolus injections on a Siemens ECAT EXACT HR camera in 3D mode. During the first part of the PET examination persons wore eye pads and were instructed to keep their eyes closed. In order to prevent subjects from recalling images during the resting condition, they were instructed to silently count backwards in steps of 7 from a three digit number. 10 seconds prior to scanstart under activation conditions, the subjects were told the keyword and they started to recall the picture. Each condition was replicated 4 times in balanced order ABBABAAB with A = rest and B = activation. Afterwards, eye pads were removed and 4 measurements with an 8 Hz alternating checkerboard were performed. Images showing percent CBF increase between resting condition and activation conditions were calculated from average images of each condition and smoothed with a median filter of 8 mm radius. To define primary visual cortex images of checkerboard activation were thresholded at 20% CBF. This contiguous volume of interest was transfered to images during visual recall and average regional CBF change within this VOI was determined.

Results: The region of primary visual cortex did show a highly significant CBF change during activation by alternating checkerboard across all subjects but no activation during visual imagery (t-Test, = 0.05).

Conclusion: The results of our study support the findings that the primary visual cortex is not involved in visual imagery in normal subjects. Due to the individual, functional definition of the target area, here is only little uncertainty in localizing primary visual cortex, because checkerboard patterns nearly exclusively activate area V1 and V2.

1. Roland PE, et al. Trends Neurosci 1994; 17:281–7
2. Kosslyn SM, et al. J Cognit Neurosci 1995; 5:263–87
3. Cohen et al. Brain 1996; 119:89–100

## V/14

NEURAL PLASTICITY IN THE VISUAL SYSTEM OF THE CONGENITAL PERIPHERAL BLIND – MRI SUGGESTING ALTERED OPTIC PATHWAYS, BUT NORMAL OCCIPITAL CORTEX. *F. Uhl (1), M. Breitenseher (2) and L. Deecke (1). Neurologische (1) and Radiologische (2) Universitätsklinik Wien, Währinger Gürtel 18–20, A-1090 Wien*

Magnetic resonance imaging (MRI) was used to evaluate possible morphological changes of the visual system in 12 patients suffering from congenital blindness of peripheral origin.

While their optical pathways showed degeneration, hypoplasia or atrophy in 7 out of 12 persons, the occipital cortex appeared normal in all cases. This dissociation is in contrast to the assumption that visually-deprived cortex may undergo degeneration. It corroborates findings of our research group that their occipital cortex is involved in non-visual functions. E.g. their inferior-occipital SPECT blood flow is increased [Neurosci Let 150:162, 1993]. DC potentials in the EEG were enhanced occipitally during active touch/Braille reading [Neurosci Lett 124:256, 1991] and even in tactile imagery [Electroenceph Clin Neurophysiol 91:249, 1994].

## V/15

STATE DEPENDENT MODULATION OF VISUAL PROCESSING IN THE MACAQUE. *Wim Vanduffel*[1], Roger B.H. Tootell[2] and Guy A. Orban[1]. [1] Laboratorium voor Neuro- en Psychofysiology, Faculteit Geneeskunde, Kuleuven, Herestraat 49, B-3000 Leuven, Belgium; [2] Massachusetts General Hospital-NMR Center, 149 13th Street, Charlestown, MA 02129, USA*

To localize state-dependent effects in the macaque visual system, we compared the metabolic activity evoked during an orientation identification task to that of a localization task by using the double-label deoxyglucose (DG) technique. It enables us to make quantitative comparisons between two conditions in the same sections from the same animal. In the identification task, the monkeys (N = 3) had to indicate by a saccade to a right or left target point whether a circular square wave grating was tilted clock- or counterclockwise from vertical. In the localization task they had to localize a target point which was randomly presented at the left or right side of the grating. The visual stimulation and trial rate was the same in both conditions. $^3$H-DG was used to label identification related metabolic activity and $^{14}$C-DG was injected to label metabolic activity during the localization task (see Vanduffel et al., Soc. Neurosci. Abstr. 21:311, '95).Quantification of isotope uptake revealed retinotopic specific suppression of acitivity related to the identification task outside the representation of the stimulus in the LGN and striate area V1. In the extrastriate visual cortex, differential DG uptake was observed in areas MT/MST (localization > identification) and the parieto-occipital area (PO/area V6) (identification > localization). In non-visual areas, higher identification-related DG uptake was observed in the posterior and anterior cingulate areas 23 and 24, the entorhinal cortex (areas 23 and 24), the ventral putamen, and mediodorsal thalamus. On the other hand, the substantia nigra pars reticulata and the ventral globus pallidum showed higher localization-related activity.This is the first experimental evidence for state-dependent modulation of neuronal activity at the level of the LGN. Furthermore, we could identify a substantial portion of a network of interconnected visual and non-visual areas involved in a simple discrimination task.

## V/16

REPRESENTATION OF CATEGORIES OF COMPLEX IMAGES IN ANTERIOR TEMPORAL CORTEX. *R. Vogels. Laboratorium voor Neuro- en Psychofysiology, Faculteit Geneeskunde, Kuleuven, Herestraat 49, B-3000 Leuven, Belgium*

Previous behavioral results (Vogels, Soc. Neurosci. Abstr, 1994) have demonstrated that rhesus monkeys are able to categorize physically dissimilar images of complex visual stimuli in classes. In order to determine the neural representation of these ordinate-level categories, single cell recordings were made in the anterior temporal cortex of 2 rhesus monkeys during categorical discrimination of color images of trees or fish and other natural or artificial objects (nontrees; non-fish). On a trial, a single stimulus was presented during fixation and the monkeys made a leftward or rightward saccade on presentation of the tree (fish) or non-tree (non-fish) image respectively. For each session, 60 images were drawn out of a set of 407 images for each category. We have recorded from 266 stimulus selective anterior temporal cortical neurons (histologically verified in one monkey). Most of these neurons were highly stimulus selective: 50% of the neurons responding with at least one third of the maximal net response to 9 or less stimuli. Scrambling the image reduced the average response, paralleling the detoriation of the categorization behavior with increasing degree of image scrambling. Most neurons tested also showed invariance for size, position and the presence of color. In order to relate stimulus selectivity to categorical discrimination, we computed for each neuron the proportion of tree or fish images (P) relative to the number of images eliciting at least one third of the maximal net respons. 14% of the neurons responded mainly to non-trees or non-fish only (P = 0). Most other cells responded selectively to images of trees (fish) and non-trees (non-fish), but 12 % of the neurons responded mainly or exclusively to tree or fish images (P > 0.90). However each of the latter neurons showed no response to some of the images of the same behavioral category. These results argue against a prototype representation at the single cell level, but on the other hand favor exemplar-based models of categorization. The data suggest that visual, natural categories are represented by the combined response of units coding for small subsets of the class.

Supported by GSKE

## V/17

THE DEVELOPMENT OF VISUAL FUNCTIONS IN CEREBR-ALLY BLIND CHILDREN AFTER SYSTEMATC TRAINING. *R. Werth, S.F. Bucher, K. Seelos. Institute for Social Pediatrics, Heiglhofstr. 63, D-81377 München, Department of Neurology, Marchioninistr. 23, D-81377 München, Department of Neuroradiology, Marchioninistr. 23, D-81377 München*

It was investigated whether visual functions can develop in children who were blind due to traumatic cerebral lesions or due to cerebral asphyxia when the visual field is systematically stimulated with light.The visual field of 6 children (aged between 1 and 15 years), who were blind due to a traumatic cerebral lesion and the visual field of 17 children (aged between 1 and 7 years) who were blind after cerebral asphyxia, was systematically stimulated every day for at least 10 minutes with light spots. Before the beginning of this visual field training, blindness had persisied for at least one year without a sign of spontaneous recovery. The extension of the funcional visual fields and functional luminance difference thresholds were measured with a specially designed arc-perimeter. Recovery of visual functions was found in 3 children with cerebral trauma and in 12 asphyxic children within a training period of 12 weeks. The children of a control group (10 suffering from traumatic cerebral lesions and 38 suffering from perinatal asphyxia) who did not receive visual field training, did not recover within one year. It was further investigated whether brain tissue in the area of the damaged striate cortex of two children, who recovered from blindness, could still be activated by light stimuli, using functional magnetic resonance imaging. In both children, vital brain tissue in the area of the visual cortex contralateral to the blind visual hemifield was activated by light. Spared tissue in the striate and extrastriate visual cortex and underlying white matter may be the anatomical basis of recovery of vision in children suffering from cerebral blindness.

## V/18

ERP INDICANTS OF FRONTAL AND OCCIPITAL BRAIN MECHANISMS MEDIATING SPATIALLY DIRECTED VISUAL PROCESSING. *A. Zani & A. M. Proverbio, Istituto di Psicologia, National Research Council, Viale Marx 15, 00137 Rome, Italy; Department of Psychology, University of Trieste, Via dell'Universita' 7, 34123 Trieste, Italy*

Functional neuroimaging and neurophysiological studie indicate the existence of a posterior attention system strongly connected with dorsolateral frontal areas involved in the control of spatially directed visual processing. The goal of the present study was to investigate the spatio-temporal dynamics of attention control mechanisms by means of high density electrophysiological recordings in healthy subjects during selective attention tasks.
ERPs were recorded from 30 scalp sites and 2 bipolar ocular montages in 10 young paid right-handers. Averaged ears served as reference. Stimuli were 4 isoluminant black and white gratings of 1 and 7 cpd randomly presented at 2 eccentric locations of the upper fields. The task consisted in selectively attending and responding to a given conjunction of frequency and location, while ignoring all other stimuli. Both accuracy and speed of response were emphasized. ERP waveforms were computed for each stimulus, attention condition and electrode site. Mean area and peak latency and amplitude values were computed for each ERP component of interest, and underwent to repeated measures analyses of variance. Laplacian Scalp Current density (SCD) maps based on thespherical spline mo-

del were also computed to localize the principal source of attention effects.
Results showed an early attention modulation of sensory responses at the *primary visual areas* as early as 60–80 msec as a function of frequency selection, location selection, and the conjunction of both. Very interestingly, frontal areas also showed a strong involvement in attentional control as well as a specific activation related to the gating of irrelevant information. This was reflected at the scalp as a strong positive focus over *dorso-lateral pre-frontal areas* in between 180–230 msec post-stimulus latency in response to irrelevant stimuli. Some conclusions can be drawn on the basis of these data. First, they indicate a role of executive control for anterior brain region in visual sensory gating by the *suppression* of sensory-perceptual processing of irrelevant inputs. Second, they show that occipital areas are involved in the attentional selection of visual information starting at an early latency level.

## V/19

OCULOMOTOR SCANNING PATTERNS IN PATIENTS WITH PARIETAL OR FRONTAL LOBE DAMAGE. *J. Zihl[1] and N. Hebel. Max-Planck-Institut für Psychiatrie, Klinisches Institut, Kraepelinstraße 10, D-80804 München, and [1] Ludwig-Maximilians-Universität, Institut für Psychologie, Neuropsychologie, München, Germany*

Eye movements were recorded during the inspection of dot patterns in control subjects (including normal subjects and patients with closed head trauma without morphological brain damage) and in patients with acquired unilateral brain damage involving posterior parietal or frontal cortical regions. Control subjects adapted their oculomotor scanning pattern effectively to the stimulus configuration. Patients' oculomotor scanning patterns were characterized by a rather rigid sequence of fixations and saccades, with no evidence of a systematic and flexible spatio-temporal organization. In a stimulus condition where dots were grouped, patients with frontal damage accurately shifted their gaze between the dot groups, but had difficulties with dot sampling. In contrast, patients with posterior parietal damage were unable to shift their eyes accurately between dot groups. These observations suggest that posterior parietal damage mainly affected the visuo-spatial guidance of the scanpath, while frontal damage impaired its planning. It is concluded that both posterior parietal and frontal brain structures and their reciprocal connections are part of a distributed neural network subserving visually-guided oculomotor scanning, and that the spatio-temporal organization of the scanpath critically depends on both structures and their close cooperation.

## V/20

THE EFFECT OF LIGHTING CONDITIONS ON THE JUDGEMENT OF THE SUBJECTIVE VISUAL VERTICAL IN PATIENTS WITH PARIETAL LESIONS: *C. Zoelch, G. Kerkhoff: EKN – Clinical Neuropsychology Research Group, Krankenhaus Bogenhausen, Dachauerstr. 164, D-80992 München, Germany*

Patients with parietal lesions often show disturbances in visuospatial perception. Little is known about possible modulatory effects of context or lighting on spatial perception in these patients. We investigated if different lighting conditions modulate the perception of the Subjective Visual Vertical (SVV) in patients with parietal lesions.
Thirteen patients with right hemispheric vascular lesions centering on the parietal lobe and 36 normal controls were tested in a PC-based measurement of the SVV. Each subject performed ten trials, five with a clockwise and five with a counterclockwise rotation towards the objective vertical (step-width 0.5, size of the bar: 16 cm × 1.4 cm), while sitting with the head and body oriented earth-vertical. The PC-monitor was covered with an ovale black mask to eliminate the margins of the screen as visual reference. Subjects were tested under two conditions: 1) normal lighting (400 lux illumination) which allowed the subject to see contours of the testing

chamber, vs. 2) total darkness (0,1 lux) where only the illuminated bar is visible on the PC-monitor. No visual contours were visible to the subjects in this condition. The average judgment of the SVV over all trials in the two lighting conditions was analyzed statistically by paired t-tests.

Normal subjects showed a nearly perfect judgment of the SVV in both conditions (light: M 89.7, sd 0.8; dark: M 89.8, sd 0.9). The patients with parietal lesions showed significant, counterclockwise deviations from the objective vertical (90) in both conditions: (light: M 92.8, sd 1.7; dark: M 92.8, sd 2.6). Their performance in the 'dark' condition was significantly inferior to that in the 'light' condition ($t = -3.51$, $df = 12$, two-tailed $p < 0.004$).

We conclude that right parietal lesions in human subjects cause a counterclockwise rotation of the SVV. The removal of visual contours which can serve as a reference (such as contours of the testing chamber) in total darkness incresed this deficit significantly in the patient group. In contrast, normal subjects easily compensated for the lack of visual contours in darkness during judgement of the SVV.

# Author Index

Aase, H   DP/17
Ábrahám, I   DP/9, MO/1
Adan, RAH   Sym6/2
Ammassari-Teule, M   Sym1/3, Sym12/2
Ángyán, L   MO/2
Anokhin, PK   LM/25
Armstrong-James, M   Sym14/1
Arsic, S   DP/1
Artigas, J   Sym3/3
Arzi, M   V/9
Aschauer, HN   ES/3
Aschersleben, G   MO/7, T/1
Assal, G   V/1
Azzoni, A   LS/11
Baamonde, C   DP/4, LM/16
Barash, S   Sym2/3
Barea-Rodriguez, E J   LM/12, LM/17
Barras, C   Sym10/4
Baune, A   MO/3, MO/6
Bellmann, A   V/1
Berger, DF   DP/17
Berridge, KC   MO/4
Bertolasi, L   MO/10
Beyer, B   T/6
Birbaumer, N   MO/9
Bisiach, E   Sym8/1
Bley, M   DP/11
Boisson, D   N/1, V/9
Bokonjic, D   DP/14, ES/7, ES/9
Bonnet, M   T/5
Bovet, P   DP/6
Bozovic, Z   DP/15
Bramwell, DI   Sym3/3
Brandner, C   LM/1
Brandt, Th   Sym7/1
Breitenseher, M   V/14
Bremmer, F   Sym2/2
Brown, CM   Sym9/1
Brugger, P   DP/2, FN/6, LM/2
Bucher, SF   Sym7/1, V/17
Bures, J   Inv/1, V/10
Calabrese, P   LM/3
Calcagni, ML   DP/3
Caminiti, R   Sym2/1
Cappa, A   DP/3
Car, M   PP/4
Carey, DP   V/2
Casini, L   T/5
Castellano, C   LM/4
Cekic, S   ES/6, ES/8
Cerutti, DT   V/4
Cestari, V   LM/4
Ciamei, A   LM/4
Cieza, A   T/6
Cioni, G   Sym13/1
Clarke, C   DP/7
Clarke, S   FN/1, Sym3/4, V/1
Clayton, NS   Sym10/1
Coenen, VA   LM/11
Cohen, R   LS/5
Cole, T J   PP/2
Colleypriest, BJ   LM/5
Colombo, G   MO/8
Colombo, MR   Sym8/1
Cools, AR   ES/10
Corda, MG   ES/4
Cornelissen, PL   LS/1

Cottier Eskenasy, A-C   FN/1
Cowey, A   LM/29, Sym3/2
Cramon, DY von   LM/3
Crawford, CA   MO/4
Cromwell, HC   MO/4
Crusio, WE   LM/15, Sym1/1
Curt, A   MO/8
Daniele, A   LS/2
Danysz, W   Sym12/1
Daprati, E   MO/5
Davies, DC   ES/1
De Cock, P   V/11
De Jong, GI   DP/9
De Rossi, G   DP/3
De Souza Silva, M A   LM/7
DeSimone, A   FN/7
Deecke, L   T/11, V/14
Deger, K   LS/3
Delatour, B   LM/6
Demonet, J-F   Sym9/2
Deruelle, C   Sym13/3
Di Betta, AM   LS/2
Di Chiara, G   ES/4
Di Virgilio, G   FN/1
Dicke, PW   Sym2/3
Dicke, P   Sym7/4
Dienel, A   T/6
Dierssen, M   DP/4, LM/16
Dieterich, M   Sym7/1
Dietz, V   MO/8
Dijkerman, H C   V/2
Dobel, C   LS/5
Dobric, S   ES/7
Dobric, S   ES/9
Drago, J   MO/4
Dreszer, J   T/10
Driver, J   Sym8/2
Duhamel, J-R   Sym2/2
Ebenbichler   T/11
Ebner, K   ES/2
Ehrenstein, WH   T/4, V/3
Eisner, W   LS/9
El Bab, MF   LM/5
Engelmann, M   ES/2, ES/14
Erb, M   MO/3, MO/6
Escorihuela, RM   LM/16
Fabre-Grenet, M   Sym13/3
Farnè A   N/1
Faure, JM   ES/1
Ferber, S   N/2
Fernández-Teruel, A   LM/16
Ferrara, A   T/5
Fetter, M   N/4
Feucht, M   DP/5
Fey, J   LM/12, LM/17
Filippini, V   LS/2
Fink, H   LM/24, Sym6/1
Fischer, MH   T/3
Fletcher, PC   Sym4/1
Florez, J   DP/4
Flórez, J   LM/16
Franck, N   MO/5
Frankiewicz, T   Sym12/1
Freund, HJ   Sym5/2
Freund, H-J   MO/17
Friederici, AD   LS/8, Sym9/3
Frith, C   Inv/2

Froelich, L   DP/19
Fuchs, K   ES/3
Fuster, JM   Inv/3
Füreder, T   ES/3
Gaffan, D   Sym10/2
Gainotti, G   DP/3, LS/2, LS/11
Ganz, RE   T/4
Gasparini, F   LS/11
Gaymard, B   Sym7/2
Gec, V   LS/6
Gehrke, J   MO/7
Geigenberger, A   LS/7
Gelder, B de   DP/6
Geminiani, G   Sym8/1
Georgieff, N   MO/5
Georgiev, V   ES/12
Gerfen, CR   Sym6/3
Gerhard, E   ES/3
Gerhardt, P   LM/24
Gerlai, R   Sym1/2
Gisquet-Verrier, P   LM/6
Goddard, P   DP/7
Goebel, R   V/6
Goessler, R   DP/5
Goetz, I   DP/16
Goldenberg, G   N/8
Gonzalez, MI   ES/1
Grabowska, A   V/12
Graves, RE   DP/2, LM/2
Grodd, W   MO/3, MO/6, MO/9
Gruden, MA   DP/8
Haan, E de   LM/21
Habs, D   T/6
Haers, M   Sym13/5
Hagoort, P   Sym9/1
Hahne, A   LS/8
Hamilton, M   ES/1
Hansen, PC   LS/1
Harders, AG   LM/3
Hari, R   Sym11/1
Harkany, T   DP/9, MO/1
Harvey, M   N/3, Sym8/4
Hasenöhrl, RU   LM/7, LM/24
Havel, P   MO/16
Haxby, JV   Sym3/1
Hebel, N   V/19
Heide, W   Sym7/3
Heiss, WD   V/13
Heiss, W-D   DP/11, DP/13
Helbing, N   DP/19
Hermsdörfer, J   LM/8, N/8
Herzog, H   FN/7, Sym9/1
Hesselmann, V   FN/7
Heywood, CA   Sym3/2
Hilker, R   DP/13
Hodges, JR   Sym9/4
Hoheisel, U   FN/2
Höller, P   LM/9, LM/20
Hölter, SM   PP/1
Hornik, K   ES/3
Hout, BM van den   Sym13/5
Hübner, C   LM/10

Hurlbert, AC  Sym3/3
Huston, JP  LM/7, LM/24
Hütter, BO  LM/11
Ilg, U  Sym7/4
Ilmberger, J  LS/9
Indefrey, P  Sym9/1
Ivanus, J  DP/15, LS/6, MO/14
Jaskowski, P  MO/15
Jeannerod, M  MO/5, MO/17, N/1
Jelen, P  ES/5
Jentjens, O  LM/7
Joëls, M  PP/2
Jost, K  MO/1
Jovanova-Nesic, K  DP/14, ES/7, ,ES/9
Jürgens, U  LS/10
Kalauzi, A  DP/15
Kalbe, E  DP/11
Kaminsky, Yu  V/10
Kämpf, P  PP/3
Karbe, H  DP/11
Karcher, S  LM/19
Kardatzki, B  MO/9
Karnath, H-O  N/2, N/4, N/7, Sym8/3
Kaske, A  FN/2, FN/3
Kasper, S  ES/3
Keck, ME  MO/8
Kegel, G  T/8
Keller, I  DP/10
Keller, J  V/4
Kentridge, R W  Sym3/2
Kerkhoff, G  N/5, N/8, T/11, V/20
Kessler, J  DP/11, DP/13, V/13
Kiernan, K  DP/7
Klimaschewski, L  FN/2
Kloet, ER de  ES/13, PP/2
Klose, U  MO/3, MO/6, MO/9
Knuf, L  T/1
Köbbel, P  LS/5
Kömpf, D  MO/15, Sym7/3
Konjevic, G  PP/4
Koshikawa, N  ES/10
Kossut, M  Sym14/2
Krasemann, P  LM/9
Krause, B  FN/7
Kriz, G  LM/13
Kublik, E  FN/4
Kurbatova, LA  DP/8
Làdavas, E  N/6
Landauer, N  T/9
Landgraf, R  ES/2, ES/14, PP/3
Landis, T  LM/2, N/10, N/9
Lang, W  T/11
Langen, EG de  LS/4, T/2
Lavenex, P  Sym10/4
Leblanc, P  LM/14
Lecca, D  ES/4
Ledenboer, A  ES/13
Leifert, G  T/6
Lejeune, H  T/5
LeMare, C  LM/29
Leonard, BE  DP/9
Leonards, U  V/5
Levine, MS  MO/4
Liegeois, F  Sym13/3
Lippert-Grüner, M  DP/12
Lotze, M  MO/9
Luiten, PGM  DP/9, MO/1
Lumbreras, M  DP/4
Maarouf, LF  LM/15
Macar, F  T/5
Maeder, P  V/1
Makashvili, M  FN/5

Mancini, J  Sym13/3
Maquet, P  T/5
Maravita, A  MO/10
Marican, CCG  LM/15
Markowitsch, HJ  LM/3, V/13
Marquardt, C  LM/8
Marra, C  LS/11
Martínez-Cué, C  DP/4, LM/16
Martin, P  LM/19
Martinez, JL Jr  LM/12, LM/17
Marzi, CA  MO/10
Mathis, C  Sym12/4
Maurer, K  DP/19
Mayer, E  N/9
Mecklinger, A  Sym9/3
Melan, C  Sym12/4
Mele, A  Sym12/2
Mense, S  FN/2
Merzenich, MM  Sym11/2
Meszaros, K  ES/3
Meuli, R  V/1
Meyer, OC  PP/2
Mielke, R  DP/13
Miller, H  N/3
Mills, AD  ES/1
Milner, AD  MO/13, Sym8/4, V/2
Miniussi, C  MO/10
Minniti, A  V/7
Mirenowicz, J  LM/28
Montoya, P  MO/9
Muckli, L  V/6
Müller, Stroh  MO/11
Müller, U  LM/18
Munk, MHJ  Sym11/3
Musia, P  FN/4
Müsseler, J  T/1
Nase, G  V/5
Nedovic, G  DP/15, MO/14
Nesic, M  ES/6
Nesic, V  ES/6
Niemann, K  LM/11
Niemeier, M  N/4, N/7
Nieuwenhuizen, O van  Sym13/5
Nowacka, A  FN/8
Nowak, D  LM/8
Nyakas, C  DP/9, MO/1
Oitzl, MS  ES/13, PP/2
Oliverio, A  Sym12/2
Òobolin, VA  DP/8
Orban, GA  V/15
Ostojic, Z  PP/4
Palomares, A  Sym3/3
Parnas, J  DP/6
Parsons, CG  Sym12/1
Pascalis, O  Sym13/3
Patterson, K  Sym9/4
Paulesu, E  Sym4/2
Paulignan, Y  MO/17, N/1
Pavani, F  N/6
Penke, B  DP/9, MO/1
Peper, M  LM/19
Perenin, M-T  V/9
Peric, P  ES/8
Philipp, J  LM/8
Pijnappels, M  MO/8
Piras, G  ES/4
Piscevic, V  MO/14
Pisella, L  V/9
Pletnicov, MV  LM/25
Popelier, T  DP/6
Popovic, M  DP/14, ES/7, ES/9
Popovic, N  DP/14, ES/7, ES/9

Pöppel, E  LM/10, T/10, T/6
Pörtner, L  LM/20
Postma, A  LM/21
Poucet, B  Sym10/3
Pouthas, V  T/5
Preilowski, B  MO/12
Prinz, W  MO/7, T/1
Pritchard, CL  MO/13
Privou, C  LM/24
Proshin, AT  LM/25
Proverbio, AM  V/7, V/18
Przybyslawski, J  LM/23
Rajsic, N  DP/15
Rakic, Lj  DP/14, PP/4
Rapaic, D  DP/15, MO/14
Ravens-Sieberer, U  T/6
Regard, M  FN/6, LM/2
Regli, L  V/1
Régnier, C  V/9
Regolin, L  LM/22
Reinert, A  FN/2
Reinshagen, G  LM/19
Rensink, AAM  DP/9, MO/1
Réthy, S  V/8
Rettenbach, R  V/5
Reulen, H-J  LS/9
Reverdin, A  N/9
Rex, A  Sym6/1
Reymann, KG  Sym12/3
Rizzolatti, G  Sym5/1
Rochat, P  Sym13/2
Rockstroh, B  LS/5
Rode, G  N/1, V/9
Roehrenbach, C  FN/6
Romero, B  DP/16
Rossetti, Y  N/1, V/9
Rossi-Arnaud, C  Sym1/3
Rossier, J  V/10
Rothwell, J  MO/10
Roubertoux, PL  LM/15
Roullet, P  LM/23, Sym12/2
Rowan, A  N/3
Roy, A  N/1
Rugg, MD  Sym4/3
Rümbeli, M  FN/6
Ruzdijic, S  PP/4
Rymarczyk, K  T/10
Saar, J  LM/19
Sabel, BA  V/4
Sagvolden, T  DP/17
Saigusa, T  ES/10
Samadashvili, I  FN/5
Sanes, J  MO/10
Sara, SJ  LM/23
Sasvári, M  DP/9, MO/1
Schellig, P  MO/12
Schenk, F  LM/1
Schenk, F  Sym10/4, V/10
Schildein, S  LM/24
Schmid, UD  LS/9
Schmids, SW  PP/2
Schmidt, U  LM/9, LM/20
Schonen, S de  Sym13/3
Schönle, PW  LS/5
Schubert, M  MO/8
Schultz, W  LM/28
Schütz, G  PP/2
Schwarz, C  Sym2/3
Schwarz, M  LS/7
Schwegler, H  Sym1/1
Schwender, D  LM/10
Sedgwick, EM  LM/5

Seelos, K   V/17
Seitz, R   Sym9/1
Shayi, S   FN/5
Sherstnev, VV   DP/8, LM/25
Shumova, EA   DP/8
Siegel, J   ES/11
Sieghart, W   ES/3
Sillaber, I   PP/3
Silveri, MC   LS/2
Simminger, D   DP/19
Singer, W   V/6
Sireteanu, R   V/5
Slater, A   Sym13/4
Sluyter, F   LM/15
Soffié, M   LM/14, LM/27
Sommer, FT   MO/3, MO/6
Spanagel, R   PP/1, PP/3
Specht, C   FN/7
Stankovic, M   ES/6
Staub, F   V/1
Steffen, A   T/7, T/8
Stein, BE   Inv/4
Stein, JF   LS/1
Steinbüchel, N v   Sym 11/4, T/6, T/9, T/10, T/13
Steiner, H   Sym6/3
Steinlein, OK   Sym1/4
Steinmann, C   DP/16
Stiers, P   Sym13/5, V/11
Stoltysik, S   ES/5
Stompe, T   ES/3
Storogeva, ZI   LM/25
Strasburger, H   V/4
Struppler, A   MO/16
Studener, R   DP/5
Sturm, W   FN/7
Sujic, RD   DP/18
Suljagic, S   DP/15
Szatkowska, I   V/12

Szelag, E   T/10
Tallal, P   Inv/5
Tchekalarova, J   ES/12
Terhaag, D   DP/12
Theml, T   DP/16
Thiel, A   V/13
Thier, P   Sym2/3, Sym7/4
Timmerman, W   MO/1
Tobeña, A   LM/16
Tokarski, J   FN/8
Tomaz, C   LM/7
Tommasi, L   LM/26
Tootell, RBH   V/15
Tribolet, N de   V/1
Trojniar, W   FN/8
Tuinstra, T   ES/10
Turner, R   Sym4/4
Ugresic, N   ES/7
Uhl, F   DP/5, N/8, T/11, V/14
Ungerer, A   Sym12/4
Valenza, N   N/9
Vallar, G   Sym2/4
Vallina, IF   DP/4, LM/16
Vallortigara, G   LM/22, LM/26
Van Waas, M   LM/27
Vandenbussche, E   Sym13/5, V/11
Vanduffel, W   V/15
Varga, J   DP/9, MO/1
Venneri, A   T/12
Verleger, R   MO/15
Veskov, R   PP/4
Vidal, F   T/5
Villa, G   DP/3
Vital-Durand, F   Sym13/6
Voelkl, S   DP/5
Vogels, R   V/16
Voigt, J-P   Sym6/1
Voits, M   Sym6/1
Vranjes, DO   DP/18

Vries, LS de   Sym13/5
Vroomen, J   DP/6
Vuckovic, V   ES/6
Vuilleumier, P   N/9, N/10
Vukovic, MG   DP/18
Waelti, P   LM/28
Walsh, V   Sym3/4
Wascher, E   MO/15
Wauschkuhn, B   MO/15
Weber, B   DP/19
Weber-Luxenburger, G   DP/13
Weiller, C   Sym4/5
Weiskrantz, L   LM/29
Weiss, P   MO/17
Welker, E   Sym14/3
Werani, A   LS/9, T/8
Werth, R   V/17
Wessel, K   MO/15
Wildgruber, D   MO/3, MO/6
Willmes, K   FN/7
Wist, ER   MO/11
Wittmann, M   T/9, T/13
Wohlfarth, R   LM/19
Workel, JO   ES/13
Wotjak, CT   ES/2, ES/14
Wróbel, A   FN/4, Sym14/4
Zagrodzka, J   ES/5
Zajaczkowski, W   Sym12/1
Zanette, G   MO/10
Zani, A   V/7, V/18
Zarándi, M   DP/9, MO/1
Zeiner, P   DP/17
Ziegler, W   LS/3, LS/7, LS/12
Zieglgänsberger, W   PP/1
Zierdt, A   LM/8
Zihl, J   V/19
Zobel, E   LS/5
Zoelch, C   N/5, V/20
Zwirner, P   LS/10